Advances in Computer Vision and Pattern Recognition

Founding editor

Sameer Singh, Rail Vision, Castle Donington, UK

Series editor

Sing Bing Kang, Microsoft Research, Redmond, WA, USA

Advisory Board

Horst Bischof, Graz University of Technology, Austria
Richard Bowden, University of Surrey, Guildford, UK
Sven Dickinson, University of Toronto, ON, Canada
Jiaya Jia, The Chinese University of Hong Kong, Hong Kong
Kyoung Mu Lee, Seoul National University, South Korea
Yoichi Sato, The University of Tokyo, Japan
Bernt Schiele, Max Planck Institute for Computer Science, Saarbrücken, Germany
Stan Sclaroff, Boston University, MA, USA

More information about this series at http://www.springer.com/series/4205

Yongqiang Zhao · Chen Yi
Seong G. Kong · Quan Pan
Yongmei Cheng

Multi-band Polarization Imaging and Applications

National Defense Industry Press

Yongqiang Zhao
Northwestern Polytechnical University
Xi'an
China

Chen Yi
Northwestern Polytechnical University
Xi'an
China

Seong G. Kong
Sejong University
Seoul
Korea, Republic of (South Korea)

Quan Pan
Northwestern Polytechnical University
Xi'an
China

Yongmei Cheng
Northwestern Polytechnical University
Xi'an
China

This work is supported by the National Natural Science Foundation of China (61371152, 61071172, 60602056) and NSFC-NRF Scientific Cooperation Program (6151101013).

ISSN 2191-6586   ISSN 2191-6594   (electronic)
Advances in Computer Vision and Pattern Recognition
ISBN 978-3-662-49371-7   ISBN 978-3-662-49373-1   (eBook)
DOI 10.1007/978-3-662-49373-1

Jointly published with National Defense Industry Press, Beijing

Library of Congress Control Number: 2016931838

Printed on acid-free paper

This Springer imprint is published by SpringerNature
The registered company is Springer-Verlag GmbH Berlin Heidelberg

Preface

As a combination of polarization and spectral imaging, multi-band polarization imaging has demonstrated a strong potential in a wide variety of civilian and military applications. Conventional imaging devices separately provide spectral information, intensity, and surface characteristics of an object. Multi-band polarization imaging acquires simultaneously the features in reflection, scattering, and transmission, which are beneficial for the detection and identification of an object in challenging environments including concealed, camouflaged, and non-cooperative targets. This book focuses on the discovery and practical applications of multi-band polarization imaging based on the research that the authors have accomplished since 2001. Topics include multi-band polarization remote sensing systems, analysis of multi-band polarization characteristics, effects of camouflage materials on polarization bidirectional reflectance distribution function, and applications of multi-band polarization imaging in target detection.

This book is composed of seven chapters, beginning with an introduction to fundamental theories, state-of-the-art developments, and applications of multi-band polarization imaging. Chapters 2–3 describe electromagnetism, polarization analysis, principles of multi-band polarization detection, and the imaging systems designed by the authors. The methods to reduce polarization measurement errors as well as calibration methods are also described in detail. Chapter 4 contains principles of polarization bidirectional reflectance distribution, the measurement system and the P-BRDF model of coated objects and the background. In Chap. 5, target detection with multi-band polarization imaging is discussed. Various fusion detection methods are compared with theoretical analysis and experimental results. Chapter 6 demonstrates bio-inspired multi-band polarization vision systems and a data restoration technique to expand the field of view. 3D reconstruction and polarization dehazing methods are described in Chap. 7.

The authors are grateful to the Talent Zone Program of School of Automation, Northwestern Polytechnical University for their support of international collaboration.

July 2015   Yongqiang Zhao
                                                                                        Chen Yi
                                                                                Seong G. Kong
                                                                                      Quan Pan
                                                                              Yongmei Cheng

Contents

Chapter 1
Introduction

In remote sensing, imaging instruments measure the energy of the light reflected or radiated from an object of interest in a number of different wavelength bands to observe physical as well as chemical characteristics of the object such as material composition and surface temperature distribution [1]. An object in the atmosphere or on the ground provides unique spectral and polarization information during the process of interaction with the light such as reflection, scattering, and transmission. Electromagnetic characteristics of the light from an object involve the light intensity as a function of wavelength, transmission direction, and the plane of polarization. Figure 1.1 illustrates the ways that object properties are described using four basic features of electromagnetic radiation: wavelength, transmission direction, intensity, and polarization plane. Wavelength or frequency describes dynamic characteristics of an object as well as material compositions along with amplitude information. Polarization plane describes surface characteristics of the object. Laws concerning how polarization direction changes with wavelength help to characterize object surface. Polarization and spectral characteristics provide critical information, which may not be easily acquired using traditional sensing methods, to detect and identify an object, especially non-cooperative targets, often hidden or camouflaged [2, 3].

Multi-band polarization imaging refers to a combination of polarization imaging and spectral imaging techniques in sensing and data analysis. Multi-band polarization imaging techniques attempt to acquire spatial, spectral, and polarization information simultaneously. Multi-band polarization imaging has become a promising approach in remote sensing for its great potential in civilian and military applications [4]. Figure 1.2 shows a hierarchy of multi-band polarization imaging approaches. Spectral imaging extracts useful information from wavelength and intensity signals, while polarization imaging utilizes polarization and intensity information.

© National Defense Industry Press, Beijing and Springer-Verlag Berlin Heidelberg 2016
Y. Zhao et al., *Multi-band Polarization Imaging and Applications*,
Advances in Computer Vision and Pattern Recognition,
DOI 10.1007/978-3-662-49373-1_1

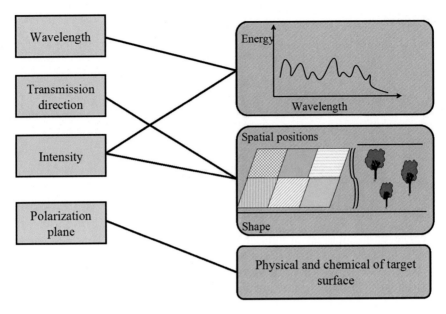

Fig. 1.1 Description of object properties in terms of basic features of electromagnetic radiation

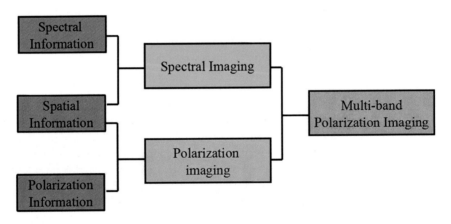

Fig. 1.2 A hierarchy of multi-band polarization imaging techniques

1.1   Spectral Imaging

1.1.1   Basic Principles

Spectral imaging, as a mixture of spectroscopy and photography, acquires spectral information of a scene at every pixel location of a digital image. Spectroscopic analysis is used to study the optical properties of an object through measuring and

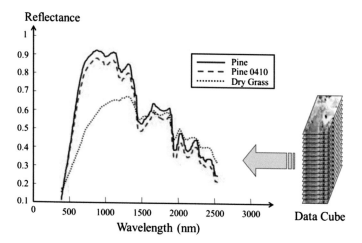

Fig. 1.3 Spectral signatures of three different pixel locations corresponding to different land covers

analyzing the electromagnetic spectrum produced by reflection and absorption. Imaging spectroscopy is used to describe spectral characteristics in the form of imaging. Various distinctions exist among spectral imaging techniques according to the criteria such as spectral range, spectral resolution, number of bands, contiguousness of spectral bands, and application. Hyperspectral imaging collects spectral information as a function of narrow and contiguous wavelength bands to completely record spectral characteristics of an object of interest [1]. Multi-spectral imaging records spectral information in a small number of bands. The measurement data in spectral imaging can be represented as a three-dimensional (3D) data cube. The data consists of multiple image slices measured at each wavelength or spectral band. The data can also be viewed as a collection of spectral information at each pixel location, which gives a vector of spectral responses at different wavelengths. The spectral values can be arranged as a form of spectral curve, where different materials demonstrate different spectral responses. Figure 1.3 shows the 3D data cube nature of spectral imaging measurements [5]. Different reflectance curves corresponding to three different pixel locations represent three different land cover types.

Spectral imaging can be characterized by four resolution types: spectral resolution, spatial resolution, radiative resolution, and time resolution [5]. Spectral resolution can be determined by the number of measured spectral bands, often a few dozens, or even hundreds of spectral bands are being used in the wavelengths that range from ultraviolet (UV) to infrared (IR) including the visible (VIS), near infrared (NIR), mid-wave infrared (MWIR), and long-wave infrared (LWIR) regions. Spatial resolution is given by the size of object that is reflected into a single pixel. Radiative resolution is defined by the minimum intensity variation of reflection or radiation of electromagnetic wave. Radiative resolution is denoted as noise equivalent reflectance and noise equivalent temperature differences in visible, near infrared bands and long infrared bands, respectively. Time resolution is defined as the time taken by an imaging spectrometer to update observation information, which is important in analyzing dynamic changes of the object.

1.1.2 Applications of Spectral Imaging Technology

Spectral features are generated from the interaction of an electromagnetic wave with the object. However, traditional remote sensing techniques have not considered such information in detection and identification. Spectral imaging techniques utilize spectral characteristics to distinguish spectral differences of an object [6]. In general, an object shows unique reflection, absorption, and radiation spectra and emission peaks in a certain spectral range. A wide spectral range should be studied to describe the characteristics of an object. Figure 1.4 demonstrates physical characteristics of various objects corresponding to different spectral ranges [5, 7]. For example, the blue light shows good transmission effects in water, which is useful to detect the depth of water. Absorbed by chlorophylls, red light has been used to identify plants. Near infrared light reveals inner characteristics of a plant leaf, which can be used to distinguish camouflaged objects from real plants. The light in the mid-wave Infrared bands can be used in smoke detection.

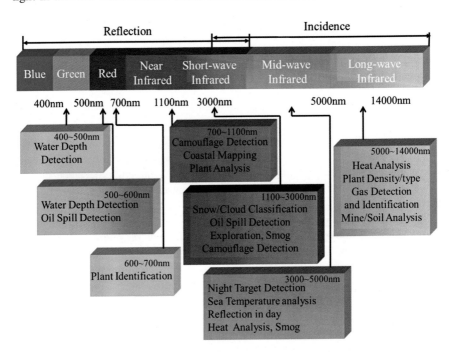

Fig. 1.4 Applications of spectral imaging technology in various spectral bands

Table 1.1 Application examples of spectral imaging

Category	Application examples
Agriculture and forestry	• Exploration and monitoring of resources [27]
	• Crop assessment [22]
	• Pest and crop disasters prediction [28]
	• Production management [22]
Military	• Camouflage identification [29]
	• Automatic target detection and recognition [2]
	• Military reconnaissance and detection [30]
Environment	• Pollution monitoring of water, air, and surface [31]
	• Hazard assessment of acid rain [32]
	• Monitoring toxic gases and wastes [31]
	• Monitoring and prediction of earthquakes and volcanos [33]
Geography	• Mineral deposit exploration [34]
	• Rock identification [34]
	• Coastal landform measurement [35]
	• Detection of shallow seabed geomorphology, plankton, water source [35]
	• Monitoring and prediction of drought and flooding [36]
Topography	• Manufacture of topographic map, photomap and thematic maps [37]
Medicine	• Diseases diagnosis [25]

There have been theoretical achievements in spectral imaging remote sensing research over last two decades, such as spectral image compression, calibration, matching, classification, identification, and information extraction. Table 1.1 summarizes application examples of spectral imaging in various categories [5, 7]:

1.2 Multi-band Polarization Imaging

1.2.1 Development of Multi-band Polarization Imaging

Polarization imaging techniques have been used in remote sensing research since 1970s. Dollfus [8] used polarization in the study of planet surface and atmospheric components. He used a non-imaging optical polarizer that can detect polarization degree of 0.1 %. Later Fraser showed that the contrast of target and background can be enhanced using polarization. Egan [9] made aerial measurements of the surface polarization characteristics. The study on polarization characteristics of the light field on the high-air platform and satellite reconnaissance system is introduced in [10], which indicates the polarization analysis in reconnaissance imaging system can evidently improve the low contrast caused by the Rayleigh scattering and Mie scattering.

Theoretical analysis estimates basic performance of a polarization imaging system. However, the above achievements did not indicate how to establish a new imaging system in the foundation of light field polarized parameters measurement, which limits the development of polarization study. Polarization imaging techniques became mature in late 1970s. Scientists developed and used polarization imagers in various remote sensing and atmospheric detection applications.

Having noticed the significance of remote sensing, NASA loaded Photopolarimeter Subsystem (PPS) and Orbital Cloud Photopolarimeter (OCPP) on the Voyager and the Pioneer Venus. Measurement of intensity and polarization of sunlight scattered by clouds and aerosols can be a very powerful tool for inferring cloud/aerosol structure and particle microphysical properties. Since 1999, the Earth Observing System (EOS) in the U.S. had installed the Earth Observing Scanning Polarimeter (EOSP), which can simultaneously measure radiation ratio and degree of linear polarization (DoLP) from 12 spectral bands in the range of 410–2250 nm. In late 1980s, French scientists developed a Polarization and Directionality of the Earth's Reflectance (POLDER) instrument that contained three polarization channels with central wavelengths of 443, 670, and 865 nm. A POLDER instrument was used to detect cloud, atmospheric aerosol, and the condition of land and ocean [11]. POLDER was first launched as a passenger instrument aboard Advanced Earth Observation Satellite (ADEOS I) on August 17, 1996. The mission ended on June 30, 1997 when communication from the host satellite failed. POLDER 2 was launched in December 2002 aboard ADEOS II. A third generation instrument was launched on board the French PARASOL microsatellite.

Advance of spectral imaging techniques enable acquisition of high spectral resolution image that can evidently improve identification capability in remote sensing. Polarization is closely related to the wavelength. Polarization images with high spectral resolution can efficiently utilize reflection, scattering, and transmission information to improve their identification capability. Multi-band polarization imaging is a combination of polarization imaging and imaging spectroscopy. Research on multi-band polarization imaging technique can be considered as a process of improving multi-band polarization imaging instrument. In a recent decade, many multi-band polarization imaging instruments had been developed, but with no achievements on standard polarization spectrometers. Multi-band polarization imaging techniques showed higher requirements, such as wider spectral range, higher spectral resolution, higher chronergy, clearer images and more details. Present polarization imaging spectrometers mainly adjust spectral bands with Liquid Crystal Tunable Filter (LCTF) and Acousto-Optic Tunable Filter (AOTF). LCTF works in intensity attenuation way while AOTF works by imposing radio wave on the birefringent crystal to set diffraction grating. AOTF changes the polarization state by rotating polarizer or adjusting phase retarder. AOTF also uses a motor to control rotation and a combination of polarizer and multiple cameras are used to acquire image data in real time. In 1999, Aerospace Laboratory of Japan designed and produced a polarization spectrometer with a motor which can acquire multi-band polarization images in the visible and infrared bands [12]. A polarization spectrometer (0.4–11.5 μm) designed by Army Research Laboratory

(ARL) in 2002 was composed of AOTF and liquid crystal variable retarder (LCVR). The polarization spectrometer was totally controlled electronically to acquire image data in real time, but with relatively low PSNR because the acousto-optic sensor was sensitive to noise. Thales Research and Technology of France studied multi-band polarization characteristics in the aspect of depolarization by combing LCTF and LCVR [13].

In mid-1980s, American researchers [9] carried out earth observation experiments on an aircraft platform, which demonstrated superiority in the measurement of polarization information in atmospheric aerosol detection and in the classification of plant and soil. Egan of New York University studied polarization characteristics using the aircrafts C-130 and B-52 in 2000 [14] and 2002 [15], respectively. The measurement of complex refractive index n of fly ash produced by Sahara in the sky of the Atlantic ocean is utilized to analyze the characteristics of n in polarized/photometric modes. The researchers of ARL indicated that different materials have unique polarization characteristics from the research on the polarization characteristics of different materials [16].

The cirrus cloud detector was a spaceborne with the function of polarization measurement, whose polarization channel can simultaneously measure linear polarization, circular polarization, and elliptic polarization. Shanghai Research Institute of Technical Physics developed in 1995 [17] and used the detector in aerial calibration experiments in 1997. Anhui Institute of Optics and Fine Mechanics developed a band tunable airborne polarization CCD principled sample machine with the support of the National Hightech R&D Program, which solved the principle technique problem of the polarization information acquisition on the airborne platform [18], first acquired aerial polarization imaging data through aerial experiments in China and certified the capability of airborne multi-band polarization information acquisition and processing. At the same time, an aerial imaging polarization detector was developed and some related flight experiments were done in 2003. Kunming Institute of Physics developed an infrared polarization spectrometer in 2007 that can obtain polarization characteristics of target radiation, which will have significant values to the detection of infrared stealth targets [19]. Universities and research institutes independently cooperated to achieve outstanding contributions in the fields on polarization reflection characteristics of single-leaf plants, medical imaging, component detection, underwater detection, and identification and polarization imager instrument.

1.2.2 Applications of Multi-band Polarization Imaging

Early polarization detection was only applied to plant soil, atmosphere detection, and nebula detection. As polarization information is divided more in detail, infrared polarization and multi-band polarization information provides more application fields for polarization information. Wolff [20] had formulated the applications of polarization imaging in 1997. Currently, primary application fields of polarization imager include atmospheric detection, earth sources investigation, medical diagnostic, military target detection and identification, imaging comprehension, and computer vision.

(A) Atmospheric Detection

According to an atmospheric polarization scattering model, the extraction of polarization states of cloud layer in different bands can help detect the distribution, type, height, and the size of cloud and atmospheric aerosol particles, where multi-band polarization detection demonstrates good identification results. Multi-band polarization detection can also be used to detect the spectrum of the natural formed metastable atomic oxygen in the upper atmosphere and measure the speed or temperature of upper atmospheric wind field, which provide useful data for atmospheric physical researchers [21].

(B) Earth Resources Investigation

A polarization spectrometer produced by National Aerospace Laboratory of Japan was first used in crop and water source investigations. Evident differences in spectral information existed among different crops and water sources with different degrees of pollution. These differences can be used to study the growth of plants, estimate crop production, and water pollution [22]. Spaceborne polarization spectrometers show better identification performance for sea and polar glacier to obtain more accurate coastline [23]. Polarization was needed in spaceborne remote sensing detection to prevent instrument damage caused by strong reflections from water surface. Ocean pollution, the size of mist in the sky of sea, and the height of wave, and the speed of wind can be reflected in the radiation polarization state [24].

(C) Medical Diagnostics

Changes of cell kernel and collagen components can have effect on optical refractive index of corresponding tissues, and then the scattering light polarization state of tissue. The tissue structures such as the size of cell kernel and the amount of collagen can be inverse from the measurement and analysis of polarization states of corresponding tissues [25]. The multi-band polarization imaging technique can measure, compare and analyze the multi-band polarization characteristics of lesion tissue and normal tissues. Applying multi-band polarization imaging helps us detect cancerous melanotic nevus more promptly and precisely. Multi-band polarization detection systems, combined with X-ray and CT imaging techniques, evidently improve the reliability and accuracy of medical diagnosis.

(D) Military Target Detection

Lots of smoke and a long exhaust plume are produced in a launched missile, which can cause strong polarization of incident radiation. Therefore, polarization detection can be used to identify a launched missile. Experiments have been done to measure polarization and photometric of a group of soldiers and the surface of military vehicles with the background of sand and plants. High polarization and low reflectance military vehicles are easy to be recognized in the sand with low polarization and high reflectance. Multi-band polarization detection technique can also detect stealth and camouflaged target, which can be used as significant information source to evidently improve performance of detection and identification [9].

(E) Material Classification

Multi-band polarization remote can provide spectral radiation information. Variations of degree of polarization and polarization angle parameters are closely related to physical and chemical characteristics of the material, which make a more detailed division when there are little differences among intensity information. For example, Fresnel reflection can indirectly describe conductive characteristics and distinguish conductors and insulators. Simultaneously measuring multiple bands of Fresnel reflection coefficients can help determine dielectric constants to be used for material classification. Metals show phase delay but not insulators. Roughness of a target surface and the surface direction can also be represented using polarization parameters with spectral bands, which is one of significant basis for multi-band polarization classification [3].

(F) Image Enhancement

The visible light is reflected and scattered by the particles in the medium when travels through the atmosphere or water, which may cause blur, low contrast, and loss of information in acquired images. In general, backscattered light and reflected light are partially polarized and polarization degree of reflected light is smaller than scattered light. Polarization degree of reflected light is determined by the optical characteristics of object surface while the scattered light is related to the size of particles and the probability of collision called particle concentration. Polarization techniques can change relative ratio of reflected light and scattered light to reduce the background noise and improve image fidelity. Utilizing polarization information has shown good results on improving the PSNR in the aspect of dust elimination and infrared image enhancement. It is effective in the processing of ghosting caused by reflection and transmission on glass surface [26].

References

1. Wang P, Zhenping B, Xiangnan L (2001) Remote sensing introduction. Higher Education Press, Beijing
2. Zhao Y, Gong P, Pan Q (2008) Object detection by spectropolarimeteric imagery fusion. IEEE Trans Geosci Remote Sens 46(10):3337–3345
3. Zhao Y, Zhang L, Zhang D, Pan Q (2009) Object separation by polarimetric and spectral imagery fusion. Comput Vis Image Underst 113(8):855–866
4. Zhao Y, Peng Q, Yi C, Kong SG (2016) Multiband polarization imaging. J Sens 5985673
5. Landgrebe DA (2003) Signal theory methods in multispectral remote sensing. Willey Publishers, New York
6. Farrell MD Jr (2005) Analysis of modeling, training, and dimension reduction approaches for target detection in hyperspectral imagery. PhD Dissertation, Georgia Institute of Technology

7. Chang CI (2003) Hyperspectral imaging: techniques for spectral detection and classification. Kluwer Academic, Dordrecht
8. Dollfus A (1975) Optical polarimetry of the Galilean satellites of Jupiter. Icarus 25(3): 416–431
9. Egan WG (1985) Photometry and polarization in remote sensing. Elsevier, New York
10. Liu J (1999) Research on tunable polarization spectral imaging techniques. Anhui Institute of Optics and Fine Mechanics, Chinese Academy of Science
11. John WW, Howard ST, Mark AP (2001) Image processing and classification for the UK Remote Minefield Detection System infrared polarimetric camera. Proc SPIE 4394:139–152
12. Drouillard TF, Searcy PA, Davis SR, Uberna RJ, Herke RA, Anderson MH, Rommel SD, Anthony EB, Damiao VB (2005) Stokes polarimetry using liquid-crystal variable retarders. Application Note, Meadowlark Optics (www.meadowlark.com)
13. Mehdi A, Francois G, Philippe R, Grisard A, Lallier E, Doffi D (2004) Multispectral polarimeteric imaging with coherent illumination: towards higher image contrast. Proc SPIE 5432:134–144
14. Egan WG, Johnson WR, Whitehead VS (1991) Terrestrial polarization imagery obtained from the space shuttle: characterization and interpretation. Appl Opt 30(4):435–442
15. Egan WG, Michael JD (2002) Synthesis of optical polarization signatures of military aircraft. In: International symposium on optical science and technology, International Society for Optics and Photonics
16. Neal RM, David D, John G (2005) Laboratory results using an active wollaston polarimeter. Proc SPIE 5888:1–5
17. Zhang Z, Wang M, Wang P (1997) Cirrus detector—a special instrument for detecting atmospheric aerosol cloud. In: The 13th annual meeting of China geophysical society
18. Yang W, Pan L, Hong J (2004) Radiation calibration of multi-spectral polarized CCD. High Technol Lett 14(10):11–15
19. Zhao J (2013) Developments of polarization imaging technology. Infrared Technol 35 (12):743–750
20. Wolff LB (1997) Polarization vision: a new sensory approach to image understanding. Image Vis Comput 15:81–93
21. Frost JW, Nasraddine F, Rodriguez J, Andino I, Cairns B (2005) A handheld polarimeter for aerosol remote sensing. In: Proceedings of SPIE, vol 5888
22. Kohzo H, Hirokimi S, Hiromichi Y, Kurosaki H (2004) Application of an imaging spectropolarimeter to agro-environmental sciences. Proc SPIE 5234:639–647
23. Park JD, Jones WL, Zec J (1999) Sea ice classification using a neural network algorithm for NSCAT. In: IEEE international conference in geoscience and remote sensing symposium (IGARSS), vol 2, pp 1040–1043
24. Fujikake H, Takizawa K, Aida T, Kikuchi H, Fujii T (1998) Electrically-controllable liquid crystal polarizing filter for eliminating reflected light. Opt Rev 5(2):93–98
25. Zhao Y, Zhang L, Pan Q (2009) Spectropolarimetric imaging for pathological analysis of skin. Appl Opt 48(10):D236–D246
26. Du A (2006) Research on key technologies of polarization imaging detection. Master's thesis, Northwestern Polytechnical University
27. Bishop CA, Liu JG, Mason PJ (2011) Hyperspectral remote sensing for mineral exploration in Pulang, Yunnan province, China. Int J Remote Sens 32(9):2409–2426
28. Huang W, Lamb DW, Niu Z, Zhang Y, Wang J (2007) Identification of yellow rust in wheat using in-situ spectral reflectance measurement and airborne hyperspectral imaging. Precision Agric 8(4–5):187–197
29. Crosby F (2004) Stokes vector component versus elementary factor performance in a target detection algorithm. Proc SPIE 5432:1–11
30. Tiwari KC, Arora MK, Singh D (2011) An assessment of independent components analysis for detection of military targets from hyperspectral images. Int J Appl Earth Obs Geoinf 13 (5):730–740

31. Kefauver SC, Peñuelas J, Ustin SL (2012) Application of hyperspectral remote sensing and GIS for assessing forest health and air pollution. In: IEEE international conference on geoscience and remote sensing symposium (IGARSS), pp 3379–3382
32. Xue Z, Gao H, Liu P, Liu J (2011) Study on the effects of simulated acid rain on wheat using the spectral reflectance at canopy and leaf level. J Agro-Environ Sci 30(2):228–235
33. Kruse FA, Kim AM, Runyon SC, Carlisle SC, Clasen CC, Esterline CH, Olsen RC (2014) Multispectral, hyperspectral and LiDAR remote sensing and geographic information fusion for improved earthquake response. In: SPIE defense and security
34. Balasubramanian UR, Saravanavel J, Gunasekaran S (2013) Ore mineral discrimination using hyperspectral remote sensing—a field-based spectral analysis. Arab J Geosci 6(12):4709–4716
35. Anderson MC, Allen RG, Morse A, Kustas WP (2012) Use of landsat thermal imagery in monitoring evapotranspiration and managing water resources. Remote Sens Environ 122:50–65
36. Pulvirenti L, Chini M, Pierdicca N, Guerriero L, Ferrazzoli P (2011) Flood monitoring using multi-temporal cosmo-skymed data: image segmentation and signature interpretation. Remote Sens Environ 115(4):990–1002
37. Zhang Y (2012) Research on three-dimensional topography measurement using structured light. Master's thesis, Harbin Engineering University

Chapter 2
Polarization Imaging

Advances in sensing technology enable acquisition of a large amount of information from the physical world. Vision techniques have played a key role in information sensing, but with a limitation that most vision systems can only perceive partial information beyond the visible spectrum. For instance, one cannot perceive the information carried by a polarized light since human vision systems are not sensitive to polarization. On the other hand, however, some marine and terrestrial animals and insects demonstrate their abilities to sense and utilize polarized lights to navigate, locate, and hunt for prey in their daily activities. Polarization is a unique characteristic of transverse wave, which is the asymmetry phenomenon of vibration direction and propagation direction. Comparing with conventional image techniques, the polarization imaging technique can detect the polarization information of targets, which will be beneficial to subsequent applications, such as target detection, recognition and identification. This chapter will focus on the principle of polarization imaging and its applications, then discuss the factors that affection polarization imaging and the method to reduce errors.

2.1 Electromagnetics and Polarization

2.1.1 Maxwell's Equations

A light traveling in the free space is known to propagate in a form of a transversal electromagnetic wave, vibrating in the plane perpendicular to its direction of propagation. Polarization is a unique phenomenon of transversal waves. Most imaging systems sense an electrical field vector only, which has dominant effects, photographic and physical, on the interaction of the light and the material. This section discusses the characteristics of an electrical field vector in a polarized light. Maxwell's equations describe the properties and interrelations between the basic

© National Defense Industry Press, Beijing and Springer-Verlag Berlin Heidelberg 2016
Y. Zhao et al., *Multi-band Polarization Imaging and Applications*,
Advances in Computer Vision and Pattern Recognition,
DOI 10.1007/978-3-662-49373-1_2

fields (electric field **E** and magnetic field **B**) and the derived fields (electric displacement field **D** and magnetizing field **H**), in which the charge distribution ρ and current density **J** are considered as well [1]. Derived field is the result of interaction of a field and the material. The differential form of Maxwell's equations is given by

$$\nabla \times \mathbf{E} + \frac{\partial \mathbf{B}}{\partial t} = \mathbf{0}$$

$$\nabla \times \mathbf{H} - \frac{\partial \mathbf{D}}{\partial t} = \mathbf{J} \tag{2.1}$$

$$\nabla \cdot \mathbf{D} = \rho$$

$$\nabla \cdot \mathbf{B} = 0$$

The interaction of the fields and the material can be described using complex physical equations. These equations will only involve a constant relationship among dielectric constant ε, magnetic permeability μ, and electrical conductivity σ if they are linear, static, and isotropic. The basic fields and derived fields satisfy a constant relationship related to the wavelength at a given spatial point.

$$\mathbf{D} = \varepsilon \mathbf{E}$$

$$\mathbf{B} = \mu \mathbf{H} \tag{2.2}$$

$$\mathbf{J} = \sigma \mathbf{E}$$

Taking the curl of the curl equations, and using the curl identity, we obtain the wave equation:

$$\nabla^2 \mathbf{E} = \mu\varepsilon \frac{\partial^2 \mathbf{E}}{\partial t^2} + \mu\sigma \frac{\partial \mathbf{E}}{\partial t} \tag{2.3}$$

Similarly, the same derivation steps can be applied to the magnetic field. The electrical conductivity σ of a dielectric material is zero in optical systems, so (2.3) can be simplified as

$$\nabla^2 \mathbf{E} = \mu\varepsilon \frac{\partial^2 \mathbf{E}}{\partial t^2} \tag{2.4}$$

Equation (2.4) is a standard wave equation and the phase velocity of light in materials or the frequency is given by

$$v = \frac{1}{\sqrt{\mu\varepsilon}} \tag{2.5}$$

A monochromatic light with frequency v or angular frequency $\omega = 2\pi v$ has an electric field $\mathbf{E}(\mathbf{r}, t) = E(\mathbf{r})e^{\pm i\omega t}$ where **r** represents a vector distance from the origin to the observation point. So the wave equation can be changed in the form of Helmholtz equation.

$$\nabla^2 E + k^2 E = 0 \tag{2.6}$$

where ∇^2 denotes the Laplacian, and k the wave number, $k = \sqrt{\mu\varepsilon}\omega = 2\pi/\lambda$. Plane wave solution is used to describe the effect of polarization.

$$\mathbf{E}(\mathbf{r},t) = E_0 \exp(\pm i(\omega t - \mathbf{k}\cdot\mathbf{r}+\delta)) \tag{2.7}$$

$\mathbf{k} = k\mathbf{s}$, where $\mathbf{s}$ denotes a unit vector in the direction of plane wave propagation, and δ a constant phase reference. If (2.7) is applied to the Maxwell's equations

$$\mathbf{E} = -\sqrt{\frac{\mu}{\varepsilon}}\,\mathbf{s}\times\mathbf{H} \tag{2.8}$$

and

$$\mathbf{H} = \sqrt{\frac{\varepsilon}{\mu}}\,\mathbf{s}\times\mathbf{E} \tag{2.9}$$

Then we use $\mathbf{s}$ to obtain the inner product

$$\mathbf{H}\cdot\mathbf{s} = \mathbf{E}\cdot\mathbf{s} = 0 \tag{2.10}$$

Equation (2.10) proves that the field is transversal, which means the electrical and magnetic field vectors are located in the same plane and are perpendicular to the direction of propagation.

2.1.2 Polarization Ellipse

Since electrical field is perpendicular to the direction of propagation, the z-axis is chosen as the direction of light wave propagation in the coordinate system where electric field vector lies in the x-y plane. The relationship between the direction of electric field vector in the x-y plane and time should be investigated. The rectangular components of electric field can be seen from the general form of plane wave solution (2.7):

$$\text{Re}\{a\exp(i(\omega t - \mathbf{k}\cdot\mathbf{r}+\delta))\} = a\cos(\omega t - \mathbf{k}\cdot\mathbf{r}+\delta) \tag{2.11}$$

where Re{ } represents the real part of a complex value. So the x, y, z components
of the electric field are given by

$$E_x = a_x \cos(\omega t - \mathbf{k} \cdot \mathbf{r} + \delta_x) \tag{2.12}$$

$$E_y = a_y \sin(\omega t - \mathbf{k} \cdot \mathbf{r} + \delta_y) \tag{2.13}$$

$$E_z = 0 \tag{2.14}$$

The electric field **E** propagates through in x-y plane at any point of the space.
Removing the $\omega t - \mathbf{k} \cdot \mathbf{r}$ term in (2.12) and (2.13):

$$\left(\frac{E_x}{a_x}\right)^2 + \left(\frac{E_y}{a_y}\right)^2 - 2\frac{E_x E_y}{a_x a_y}\cos\delta = \sin^2\delta \tag{2.15}$$

where $\delta = \delta_y - \delta_x$. The trace becomes an ellipse by rotating (E_x, E_y) with the angle θ,

$$\tan 2\theta = \frac{2a_x a_y}{a_x^2 - a_y^2}\cos\delta \tag{2.16}$$

which is called a polarization ellipse as shown in Fig. 2.1. The curve is drawn by
the tip of electric field vector on the plane which is perpendicular to the direction of
wave propagation. The principal axis has the same coordinate axis as (E_x, E_y) at
$\theta = 0$ or at an odd integral multiples of 2π. The ellipse can be rotated both in
clockwise and counter-clockwise directions. Polarization direction is called
R-polarization when the ellipse rotates in the clockwise rotation ($\sin \delta > 0$),
otherwise it is L-polarization. The state of polarization is defined by the ratio of
ellipse axis, rotation angle θ and R-polarization or L-polarization of the ellipse. Two
special cases in the polarization ellipse are:

Fig. 2.1 Polarization ellipse

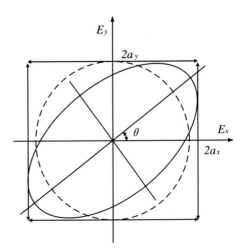

(1) If θ is integral multiple of π, (2.15) becomes a linear equation

$$E_y = \pm \frac{a_y}{a_x} E_x \qquad (2.17)$$

(2) If $a_x = a_y = a$ and δ is an odd integral multiple of 2π, (2.15) becomes a circle

$$(E_x)^2 + (E_y)^2 = a^2 \qquad (2.18)$$

Then the light is circularly polarized whose ratio of ellipse axis is 1 for an arbitrary rotation angle θ.

2.1.3 Fresnel's Equations

To obtain polarization state of an optical system, the effect of electric field at the boundary of the media of different refractivity should be calculated. In general, a beam is refracted and the other portion is reflected when a plane wave comes across a boundary of different media. Suppose that the light will not be absorbed by the medium. Fresnel's equations describe reflection and refraction. Laws of reflection and refraction prove that an incident light, reflected light, refracted light, and the normal of interface are in the same plane. We can get the same results as geometrical optics by solving Maxwell's equations and adding necessary boundary conditions: incident and reflection angles are the same but in the opposite sign. The refraction angle is given by Snell's Law. The plane consisting of incident light propagation vector $\mathbf{S}$, incident light, a reflected light, and the normal component is called incident plane as shown in Fig. 2.2.

The electric field can be decomposed into two components, parallel and perpendicular to the incident plane. For incident angle θ_i, refraction angle θ_t, amplitude

Fig. 2.2 Reflected light, refracted light, and the normal component in the incident plane

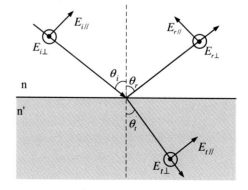

ratio r of reflection light and incident light, and amplitude ratio t of refraction light and incident light, the Fresnel's equations are given by

$$r_\perp = \frac{n\cos\theta_i - n'\cos\theta_t}{n\cos\theta_i + n'\cos\theta_t} \tag{2.19}$$

$$r_\| = \frac{n'\cos\theta_i - n\cos\theta_t}{n'\cos\theta_i + n\cos\theta_t} \tag{2.20}$$

$$t_\perp = \frac{2n\cos\theta_i}{n\cos\theta_i + n'\cos\theta_t} \tag{2.21}$$

$$t_\| = \frac{2n\cos\theta_i}{n'\cos\theta_i + n\cos\theta_t} \tag{2.22}$$

where n and n' denote refractive indices of incident medium and refractive medium, respectively. Suppose that the magnetic conductivities of the media are same as that in vacuum ($\mu = \mu' = \mu_0$). Then the reflectivity R is determined by the square of amplitude reflection coefficient:

$$R_\perp = r_\perp r_\perp^*, \quad R_\| = r_\| r_\|^* \tag{2.23}$$

and

$$|r_\perp|^2 + |t_\perp|^2 = |r_\||^2 + |t_\||^2 = 1 \tag{2.24}$$

Equation (2.24) proves that the energy of the light is conserved at the boundary of two dielectric materials.

2.2 Principles of Polarization Imaging

Polarization characteristics of a light refer to change of polarization state during transmission or reflection, which can generally be expressed by the Jones vector, Stokes vector, or Muller matrix [1, 2]. The Stokes vector was proposed when Stokes studied partially polarized light in 1852. Stokes vector represents completely as well as partially polarized light being used in polarization imaging detection. The four parameters of a Stokes vector $\mathbf{S} = [s_0, s_1, s_2, s_3]$ describe polarization information of target which can be obtained by various imaging equipment both directly and indirectly.

We commonly use electro-optic, magneto-optic, photoelastic modulation and the four detector amplitude segmentation methods to measure the Stokes vector. These

Fig. 2.3 The principle of
polarization imager based on
modulation

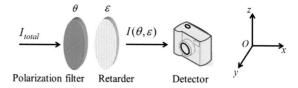

modulation methods are usually adopted as follows: Incident light gets through a
series of detecting elements (e.g. the polarizer and the phase retarder) and changes
the state of detecting elements (e.g. the polarizing angle or the phase delay angle)
by mechanical rotating or continuous periodic modulation, then get a group of
intensity values of output light and use Fourier analysis to obtain four Stokes
images, so the degree of polarization, the angle of polarization and the ellipticity of
polarization are acquired by further analysis. In 1999, Azzam [20] proposed a beam
polarization measurement system without mechanical rotation and modulation,
which can approximately get four real-time Stokes parameters by amplitude division. Then we take the single detector modulation polarization imager system as an
example to introduce the acquisition of Stokes images. Figure 2.3 illustrates the
principle of a polarization imaging detection system based on modulation. The light
is assumed monochromatic for the sake of simplicity.

The intensity of the partially polarized light is denoted as I_{total} which propagates
along the direction of z axis. The average frequency of I_{total} is denoted as v. And the
electric field **E** can be divided into two orthogonal plane waves $E_x(t)$ and $E_y(t)$ in
the direction of x- and y-axis

$$E_x(t) = a_x(t)e^{i[\phi_x(t)-\omega t]} \tag{2.25}$$

$$E_y(t) = a_y(t)e^{i[\phi_y(t)-\omega t]} \tag{2.26}$$

where $a_x(t)$ and $a_y(t)$ denote the amplitudes of electric field components in x- and
y-direction while $\phi_x(t)$ and $\phi_y(t)$ are the phases. $E_x(t)$ and $E_y(t)$ go through a
retarder with angle ϕ and polarizer of angle θ, then the electrical field component of
emergent light is

$$E(t; \theta, \phi) = E_x \cos \theta + E_y e^{i\phi} \sin \theta \tag{2.27}$$

So the intensity of emergent light is $I(\theta, \phi)$:

$$\begin{aligned} I(\theta, \phi) &= \langle E(t; \theta, \phi), E^*(t; \theta, \phi) \rangle \\ &= J_{xx} \cos^2 \theta + J_{yy} \sin^2 \theta + J_{xy} e^{-i\phi} \cos \theta \sin \theta + J_{yx} e^{i\phi} \sin \theta \cos \theta \end{aligned} \tag{2.28}$$

where the coherence matrix $\mathbf{J}$ is given by

$$\mathbf{J} = \begin{bmatrix} J_{xx} & J_{xy} \\ J_{yx} & J_{yy} \end{bmatrix} = \begin{bmatrix} \langle E_x, E_x^* \rangle & \langle E_x, E_y^* \rangle \\ \langle E_y, E_x^* \rangle & \langle E_y, E_x^* \rangle \end{bmatrix}$$

$$= \begin{bmatrix} a_1^2 & a_1 a_2 e^{i(\phi_1 - \phi_2)} \\ a_1 a_2 e^{-i(\phi_1 - \phi_2)} & a_2^2 \end{bmatrix} \tag{2.29}$$

Off-diagonal elements of the coherence matrix $\mathbf{J}$ are complex-valued, so the intensity value $I(\theta, \varepsilon)$ can be expressed as:

$$I(\theta, \varepsilon) = J_{xx} \cos^2 \theta + J_{yy} \sin^2 \theta + 2\sqrt{J_{xx}}\sqrt{J_{yy}} \cos \theta \sin \theta |J_{xy}| \cos(\beta_{xy} - \varepsilon) \tag{2.30}$$

where β_{xy} denotes the phase angle for J_{xy}. The four elements J_{xx}, J_{yy}, J_{xy}, and J_{yx} in coherence matrix $\mathbf{J}$ are determined by taking different angles of $\{\theta, \varepsilon\}$ for retarder and polarizer during image acquisition process. For example, if the angles of $\{0°, 0\}, \{45°, 0\}, \{90°, 0\}, \{135°, 0\}, \{45°, \pi/2\}$, and $\{135°, \pi/2\}$ are substituted for $\{\theta, \varepsilon\}$, the four elements are determined by

$$J_{xx} = I(0°, 0) \tag{2.31}$$

$$J_{yy} = I(90°, 0) \tag{2.32}$$

$$J_{xy} = \frac{1}{2}\{I(45°, 0) - I(135°, 0)\} + i\frac{1}{2}\{I(45°, \pi/2) - I(135°, \pi/2)\} \tag{2.33}$$

$$J_{yx} = \frac{1}{2}\{I(45°, 0) - I(135°, 0)\} + i\frac{1}{2}\{I(45°, \pi/2) - I(135°, \pi/2)\} \tag{2.34}$$

Equations (2.31)–(2.34) show that the light wave can be described by the four elements of the coherent matrix $\mathbf{J}$. Stokes proposed the four parameters to describe polarization properties of a light. According to the definition of Stokes parameters and (2.29) along with the expression in (2.31)–(2.34), Stokes parameter expressions are:

$$s_0 = \langle a_1^2 \rangle + \langle a_2^2 \rangle = J_{xx} + J_{yy} = I(0°, 0) + I(90°, 0) \tag{2.35}$$

$$s_1 = \langle a_1^2 \rangle - \langle a_2^2 \rangle = J_{xx} - J_{yy} = I(0°, 0) - I(90°, 0) \tag{2.36}$$

$$s_2 = 2\langle a_1 a_2 \cos \delta \rangle = J_{xy} + J_{yx} = I(45°, 0) - I(130°, 0) \tag{2.37}$$

$$s_3 = 2\langle a_1 a_2 \sin \delta \rangle = i(J_{yx} - J_{xy}) = I(45°, \pi/2) - I(135°, \pi/2) \tag{2.38}$$

where $\delta = \phi_1 - \phi_2$. Partially polarized light can be decomposed into independent, not polarized and complete, polarized light. The Stokes parameters of partially polarized light can be expressed using the cumulative Stokes parameter of independent not polarized and complete polarized light. So the Stokes vector **S** of partially polarized light is

$$\mathbf{S} = \mathbf{S}^{(1)} + \mathbf{S}^{(2)} \tag{2.39}$$

where $\mathbf{S}^{(1)}$ and $\mathbf{S}^{(2)}$ denote the Stokes vector of non-polarized and polarized light. The degree of polarization (DoP) is defined as:

$$\mathrm{DoP} = \frac{\sqrt{s_1^2 + s_2^2 + s_3^2}}{s_0} \tag{2.40}$$

And the polarization ellipticity is defined as

$$\beta = \tan^{-1} \frac{s_3}{\sqrt{s_1^2 + s_2^2}} \tag{2.41}$$

A vast majority of circularly polarized components in the range of instrument detection can be ignored, therefore, in most polarization imagers, $s_3 \equiv 0$ [2, 3]. For linearly polarized light, Stokes vectors can be changed to the degree of linear polarization (DoLP), polarization angle (Orient) and other parameters:

$$\mathrm{DoLP} = \frac{\sqrt{s_1^2 + s_2^2}}{s_0} \tag{2.42}$$

$$\mathrm{Orient} = \frac{1}{2} \tan^{-1} \left(\frac{s_2}{s_1} \right) \tag{2.43}$$

2.3 Polarization Imaging for Object Recognition

Polarization of the light provides a way to characterize physical properties of an object of different materials. A target produces polarization characteristics in the process of reflection and radiation on earth's surface as well as in the atmosphere. Therefore, polarization characteristics provide useful information for detection, tracking, and recognition of a target. Polarization gives a great benefit to improve the signal-to-clutter ratio, speed and precision of edge detection, accuracy and reliability of target recognition in the process of target detection [4, 5]. In 1980s, the researchers began research on polarization imaging in military applications, which involved detection and tracking of military vehicles, detection and recognition of

concealed and camouflaged targets, detection of shallow underground and underwater mines, submarine, and reef.

2.3.1 Object Detection and Tracking

The surface of an artificial target, on the ground or in the atmosphere, always
reflects and radiates some electromagnetic waves. We may receive electromagnetic
waves with a large degree of linear polarization of reflection and radiation in a
certain point of view. Comparing with that of artificial objects, the surface of natural
objects tends to be rough, so the degree of linear polarization is nearly zero. From
the observation of such differences, polarization imagers can improve the contrast
ratio of a target and the background. Artificial targets are often brighter than natural
objects in DoLP and Orient images, which provide a great clue for target detection.
Filippidis et al. [6] used polarization imagers in the application of preventing
antipersonnel and vehicle mine, shallow ground or underwater, and achieved good
results. Katkovaky et al. [7] applied polarization images in the detection of
man-made targets in the natural scenes, which greatly reduced required time for
detection and improved detection accuracy. Military targets under the cover of
natural objects can be detected using different polarization properties of electromagnetic waves that natural and man-made objects have. Polarization imagers are
also applied for the detection of underwater reef, surface or underwater naval
vessels, and to track man-made targets such as automobiles or aircrafts [8].

2.3.2 Edge Detection

According to the Fresnel's polarized reflection model [3, 5] (Figs. 2.4 and 2.5),
unpolarized lights have equal amplitude components in an arbitrary direction. After
the specular reflection of insulator, the electromagnetic wave perpendicular to the
incident plane $F_\perp$ (the dotted line plane in Fig. 2.5) has a much higher amplitude of
polarization component $I_\perp$ than $I_{||}$, while $I_{||}$ is the amplitude of polarization

Fig. 2.4 Light reflection on
the surface of material

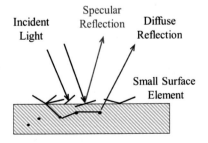

Fig. 2.5 Specular reflection

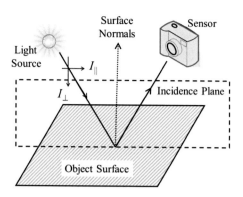

component, produced by electromagnetic wave which is parallel to the incident plane $F_{||}$. The reason for this difference is that the reflection coefficient on plane $F_\perp$ is higher. While the polarization components of electromagnetic wave have the relationship that $I_{||} > I_\perp$, which is reflected as a $90°$ difference of the phase angle between the surface of specular reflection and diffuse reflection in the polarization angle images. According to this difference, we can detect the edge of the surface of specular reflection and diffuse reflection. The maximum amplitude of electromagnetic wave I_{max} can be observed in the direction which is perpendicular to the plane of incidence after specular reflection, and the maximum intensity of electromagnetic wave I_{max} can be observed in the direction which is parallel to the incidence plane after diffuse reflection. If I_d and I_s represent the intensity of diffuse reflection and specular reflection components, $I_{||}$ and $I_\perp$ can be expressed in (2.44) and (2.45).

$$\frac{I_d}{2} + \left[\frac{F_\perp}{F_{||} + F_\perp}\right] I_s = I_\perp \tag{2.44}$$

$$\frac{I_d}{2} + \left[\frac{F_{||}}{F_{||} + F_\perp}\right] I_s = I_{||} \tag{2.45}$$

For a polarized light parallel to incident plane, the refractive index is given by

$$r_{||}(n_i, \eta, \theta_i, \theta_r) = \frac{E_{r||}(n_i, \eta, \theta_i, \theta_r)}{E_{i||}} = \frac{\eta \cos \theta_i - n_i \cos \theta_r}{\eta \cos \theta_i + n_i \cos \theta_r} \tag{2.46}$$

A polarized light is perpendicular to the incidence plane:

$$r_\perp(n_i, \eta, \theta_i, \theta_r) = \frac{E_{r\perp}(n_i, \eta, \theta_i, \theta_r)}{E_{i\perp}} = \frac{n_i \cos \theta_i - \eta \cos \theta_r}{n_i \cos \theta_i + \eta \cos \theta_r} \tag{2.47}$$

where θ_i denotes the incident angle and θ_r is the output angle. $E_\perp$ is the polarization component perpendicular to the incident plane and $E_\parallel$ is parallel with the incident plane. The relationship between θ_i and θ_r is determined by Snell's law.

$$n_i \sin \theta_i = c\sqrt{\left(\varepsilon - i\frac{\sigma}{\omega}\right)\mu} \sin \theta_r = (n - ik) \sin \theta_r = \eta \sin \theta_r \qquad (2.48)$$

The refraction coefficient of the air is $n_i = 1$, ε is the dielectric constant of the object, ω is the frequency of light, σ is the conductivity on the surface of object, μ is the magnetic conductivity, n is the single refractive coefficient and k is the extinction coefficient. The total intensity of incident light is defined as

$$I_{total}(\lambda, \eta, \theta_i) = I_{min}(\lambda, \eta, \theta_i) + I_{max}(\lambda, \eta, \theta_i) \qquad (2.49)$$

where

$$I_{max}(\lambda, \eta, \theta_i) = \frac{1}{2}I_d + \frac{r_\perp^2(\lambda, \eta, \theta_i)}{r_\perp^2(\lambda, \eta, \theta_i) + r_\parallel^2(\lambda, \eta, \theta_i)} I_s \qquad (2.50)$$

$$I_{min}(\lambda, \eta, \theta_i) = \frac{1}{2}I_d + \frac{r_\parallel^2(\lambda, \eta, \theta_i)}{r_\perp^2(\lambda, \eta, \theta_i) + r_\parallel^2(\lambda, \eta, \theta_i)} I_s \qquad (2.51)$$

According to this rule, we can acquire good results by detecting and extracting the surface of different insulator in the scene [3], and the polarization characteristics of the edge detection method can also be used in the scene with multiple targets.

2.3.3 Object Classification and Recognition

In general, a matter can be classified as either conductor or insulator according to its conductive properties. Following the Fresnel reflection model, the incident light is assumed to be unpolarized and its diffuse reflectance and specular reflection components are related to Fresnel reflection coefficient after the reflection on the surface of metal or insulator. While there is a direct relationship between the Fresnel reflection coefficient and the properties of a matter like the dielectric constant and magnetic constant, therefore, the conductive property of object can be indirectly determined through the Fresnel reflection coefficient, and then it can be classified as conductor or insulator [3–5]. The majority of specular reflection angles represent strong partial polarization characteristics after the reflection on the surface of insulators. With the enhancement of conductivity on the surface of material, the partial polarization characteristics will be weakened after specular reflection. For the insulator, there is only certain wavelengths of light can be absorbed after diffuse reflection, while as for the metal, nearly all of the light is absorbed and only a very small amount reflects out of surface. So the specular reflection mostly happens on metal surface that explains metal surface appears to be shiny. According to (2.24) and (2.25), the degree of linear polarization is defined as

$$
\begin{aligned}
\text{DoLP} &= \frac{I_{\max}(\lambda, \eta, \theta_i) - I_{\min}(\lambda, \eta, \theta_i)}{I_{\max}(\lambda, \eta, \theta_i) + I_{\min}(\lambda, \eta, \theta_i)} \\
&= \frac{r_\perp^2(\lambda, \eta, \theta_i) - r_\parallel^2(\lambda, \eta, \theta_i)}{r_\perp^2(\lambda, \eta, \theta_i) + r_\parallel^2(\lambda, \eta, \theta_i)} \cdot \frac{1}{1 + I_d/I_s}
\end{aligned}
\tag{2.52}
$$

And the polarization angle is

$$
\text{Orient} = \frac{1}{2}\tan^{-1}(\cos(\delta)\tan(2\alpha))
\tag{2.53}
$$

where $\tan\alpha = E_{oy}/E_{ox}$, δ is the relative phase. E_{oy} and E_{ox} are the maximum amplitude in the x- and y-axis. From (2.52) and (2.53), we know that the degree of linear polarization and polarization angle are closely related with material dielectric constant and incident angle. Figures 2.6 and 2.7 show that how the degree of polarization and polarization angle change with the incident angle on the surface of three types of materials, glass ($n = 1.89$), iron ($n - ik = 1.51 - i1.63$, $\lambda = 0.589$ μm) and copper ($n - ik = 0.82 - i5.99$, $\lambda = 0.65$ μm). DoLP and Orient changes with incident angle, but when the incident angle is fixed, they are shown obvious differences on the surface of metal and insulator [5].

According to the Fresnel's reflection model, approximate Fresnel coefficient is calculated using the maximum value $I_{\max}$ and the minimum radiation value $I_{\min}$ of

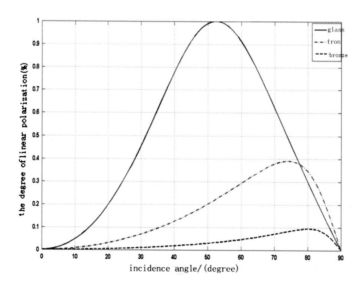

Fig. 2.6 The degree of linear polarization changes with incident angle

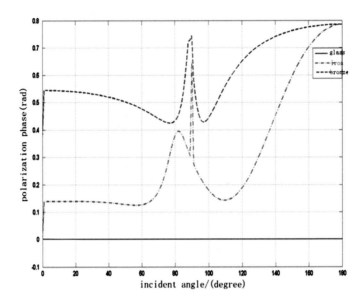

Fig. 2.7 The polarization angle changes with incident angle

the reflected light which comes out of a polarizer to obtain the information that the object surface is insulator or metal [3–5]. A sharp decrease (or increase) of the material conductivity will lead to an increase (or decrease) of I_{max}/I_{min} at a certain angle range of specular reflection, so metal or insulator can easily be determined. However, the material classification method based on the Fresnel reflection model requires mainly specular reflection and a perfect light source, otherwise it may cause a large error [5]. Since metal produces phase delay of light waves, but not insulators, another classification method uses reflected light's phase delay to metal and insulator. This method is more robust and accurate than the method using the ratio I_{max}/I_{min}, and easy to classify the objects. The polarization angle image is seriously affected by the noise. Classification of materials based on polarization characteristics has good classification performance to be widely applied to the recognition of concealed military targets, mineral substance hidden in the shallow surface or under the water, and cultural relics.

2.4 Factors Affecting Polarization Imaging

In polarization imaging detection systems, polarization parameter images usually have large errors due to the noise, non-ideal characteristics of polarization devices, and partial polarization characteristics of incident light. This section simplifies the analysis on the influence of non-ideal polarization properties and different weather conditions on the acquired polarization images.

2.4.1 Transmission of Polarized Light

Equations (2.35)–(2.37) are used to calculate Stokes parameters s_0, s_1, s_2, and s_3. $I(0°, 0)$, $I(45°, 0)$, $I(90°, 0)$, and $I(135°, 0)$ are the quantity of photoelectron, which is obtained by the beam going through the polarizer in the directions of 0°, 45°, 90°, and 135°. Figure 2.8 shows that an incident light can be decomposed into two components, parallel (P-polarized light) and perpendicular (S-polarized light) to the incident plane. P-polarized light is incident linearly polarized light with polarization direction lying in the plane of incidence. S-polarized light has polarization perpendicular to the plane of incidence. The strength of reflection from the surface is determined by the Fresnel Eqs. (2.19)–(2.22), which are different for S- and P-polarized light. Any light striking a surface at a special angle of incidence known as Brewster's angle, where the reflection coefficient for P-polarization is zero, will be reflected with only the S-polarization remaining. This principle is employed in the so-called "pile of plates polarizer" in which part of the S-polarization is removed by reflection at each Brewster angle surface, leaving only the P-polarization after transmission through many such surfaces.

The polarizer is considered having ideal transfer characteristics, $t_p = 0$ and $t_s = 1$. Since $t_p \neq 0$ and $t_s \neq 1$ in reality, however, the Stokes parameters are represented by

$$s_0' = \frac{I'(0°, 0) + I'(90°, 0)}{t_s + t_p} \tag{2.54}$$

$$s_1' = \frac{I'(0°, 0) - I'(90°, 0)}{t_s - t_p} \tag{2.55}$$

$$s_2' = \frac{I'(45°, 0) - I'(135°, 0)}{t_s - t_p} \tag{2.56}$$

Fig. 2.8 Two components of incident light

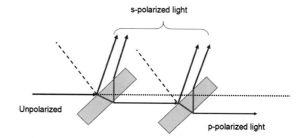

The degree of linear polarization DoLP' and the polarization azimuth angle Orient' are given by:

$$\text{DoLP}' = \frac{\sqrt{s_1'^2 + s_2'^2}}{s_0'} \tag{2.57}$$

$$\text{Orient}' = \frac{1}{2}\tan^{-1}\left(\frac{s_2'}{s_1'}\right) \tag{2.58}$$

To make a reasonable evaluation of the effect on polarization parameter image caused by polarizer's non-ideal characteristics, acquired polarization parameter images without the effect of t_p and t_s are assumed to be a noisy image. On the other hand, polarization parameter images $(s_0', s_1', s_2', \text{DoLP}', \text{and Orient}')$ with the effect of t_p and t_s are signal images. The noise-to-signal ratio (NSR) is defined as

$$\text{NSR} = \frac{\sum_{i=1}^{M}\sum_{j=1}^{N}\left(\mathrm{I}(i,j) - \hat{\mathrm{I}}(i,j)\right)^2}{\sum_{i=1}^{M}\sum_{j=1}^{N}\mathrm{I}^2(i,j)} \tag{2.59}$$

where M and N are the size of the image. In the simulation, the parameters were set to $t_p = 96\%$, $t_s = 1\%$, $M = 260$, and $N = 280$ (Fig. 2.9).

Comparing Figs. 2.10 and 2.11, the Stokes parameter images, the DoLP' image and the Orient' image are almost of no change. At the same time, we can find that the NSR value of these three images is fixed in any scenes by analyzing (2.59).

(a)   (b)   (c)   (d)

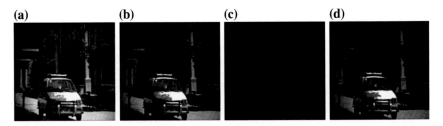

Fig. 2.9 Images collected in different polarization angles. **a** $I'(0°, 0)$, **b** $I'(45°, 0)$, **c** $I'(90°, 0)$, **d** I' (135°, 0)

(a)   (b)   (c)   (d)   (e)

Fig. 2.10 Parameter images computed using (2.35)–(2.37), (2.42), and (2.43), **a** s_0, **b** s_1, **c** s_2, **d** DoLP, **e** Orient

(a)   (b)   (c)   (d)   (e)

Fig. 2.11 Parameter images computed using (2.55)–(2.59), **a** s'_0, **b** s'_1, **c** s'_2, **d** DoLP', **e** Orient'

Table 2.1 NSR values obtained considering the effect of t_p and t_s

	s_0	s_1	s_2	DoLP	Orient
NSR	0.0309	0.0526	0.0526	0.0211	0

The NSR value of parameter images is only related to the polarizer's own characteristics and has nothing to do with the scene information. Table 2.1 shows that the NSR value of Orient' image is zero and the values of s'_0, s'_1, s'_2, and DoLP' are small, while the corresponding NSR value decreases with the improvement of permeability. Consequently, the calculation of Stokes parameter images, the DoLP' and Orient' can be done using (2.35)–(2.42), which indicates that the polarizer has ideal characteristics, and the affection of t_p and t_s should be ignored unless high accurate polarization images are needed.

2.4.2   Lighting

Under normal circumstances, the incident light is assumed to be non-polarized and will be changed to partially polarized light after reflection from the target. Non-polarized light changes into linearly polarized light after the reflection of objects with smooth surface but it will remain non-polarized when reflected from objects with rough surface. This property is used to improve the image contrast between artificial targets and natural targets in polarization images. From Sect. 2.3.2, it is easy to obtain the edge information of artificial targets in Orient images. Natural light is considered non-polarized, artificial targets can be detected using polarization imaging. In reality the sunlight becomes partially polarized after scattering through the atmosphere because of the suspended particles and water molecules in the air, which will have noise interference on the polarization images, especially Orient images, and bring more difficulty to the calibration of target and edge detection based on the polarization image. On the other hand, the polarization images may have a certain difference in different weather and light conditions because the suspended particles and water in the air vary with the weather. The degree of polarization is the strongest in clear day while there is almost no polarization in cloudy day.

2.5   Reduction of Measurement Errors

From (2.35)–(2.42), the measurement error and noise in image $I(\theta, 0)$ will transfer to the Stokes parameter images, the DoLP and Orient images and bring more errors to the subsequent target detection and recognition. How to reduce measurement error and noise of Stokes parameter images, DoLP and Orient, is the key to the success in the application of polarization imaging. Previous study demonstrated that we can combine multiple angle measurements, image pre-filtering and image fusion to reduce the effect on parameter images caused by the measurement error and noise in the image $I(\theta, 0)$ [5].

2.5.1   Multi-angle Measurement

A detailed description of Stokes parameters can be found by changing polarization angles continuously. The error caused by beam migration and the error caused by transparent inhomogeneity of light decrease as polarization angle changes. From (2.30)–(2.38),

$$I(\theta, 0) = s_0 + s_1 \cos(2\theta) + s_2 \sin(2\theta) \tag{2.60}$$

Collect N images in N polarization angles, θ_j, $j = 0, \ldots, N - 1$ from $0°$ to $360°$, then calculate the Stokes-Muller polarization parameters $\tilde{s}_0, \tilde{s}_1, \tilde{s}_2$:

$$\tilde{s}_0 = \frac{1}{N} \sum_{j=0}^{N-1} I(\theta_j, 0) \tag{2.61}$$

$$\tilde{s}_1 = \frac{2}{N} \sum_{j=0}^{N-1} I(\theta_j, 0) \cos(2\theta_j) \tag{2.62}$$

$$\tilde{s}_2 = \frac{2}{N} \sum_{j=0}^{N-1} I(\theta_j, 0) \sin(2\theta_j) \tag{2.63}$$

where $\theta_j = \frac{2\pi j}{N}$. The degree of polarization and polarization phase are:

$$\widetilde{\text{DoLP}} = \frac{\sqrt{\tilde{s}_1^2 + \tilde{s}_2^2}}{\tilde{s}_0} \tag{2.64}$$

$$\widetilde{\text{Orient}} = \frac{1}{2} \tan^{-1} \left(\frac{\tilde{s}_2}{\tilde{s}_1} \right) \tag{2.65}$$

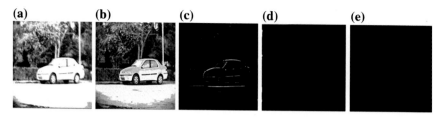

Fig. 2.12 Polarization parameter images obtained for 0°, 45°, 90°, and 135°, **a** s_0, **b** s_1, **c** s_2, **d** DoLP, **e** Orient

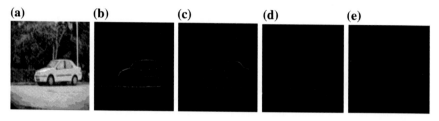

Fig. 2.13 Polarization parameter images calculated using 12 images at every 30° from 0° to 360°, **a** s_0, **b** s_1, **c** s_2, **d** DoLP, **e** Orient

Comparing (2.61)–(2.63) to (2.35)–(2.37), the Stokes parameters are only a difference of a certain constant coefficient when $\theta_j = 0°, 45°, 90°, 135°$. Figures 2.12 and 2.13 show that the Stokes parameter images, DoLP, and Orient, using (2.61)–(2.63) and (2.35)–(2.37) and twelve $I(\theta, 0)$ images calculated at every 30° from 0° to 360° (Table 2.2).

Comparing the maximum DoLP and maximum Orient calculated by the above two methods, we can find that the DoLP and Orient are maximum when $N = 12$. Analysis can be seen in Table 2.3 that the entropy of Stokes parameter images, DoLP image, and Orient image increase with multi-angle measurement. The information of every polarization parameter image increases.

Table 2.2 Maximum values of DoLP and Orient

	DoLP (%)	Orient (rad)
4-Angle measurement	78.85	2.000
12-Angle measurement	81	2.1078

Table 2.3 The entropy of Stokes parameters, DoLP, and Orient images obtained

	s_0	s_1	s_2	DoLP	Orient
4-Angle measurement	0.0218	0.0490	0.4794	0.9645	0.6619
12-Angle measurement	0.0311	1.3889	1.0470	1.5272	0.8314

2.5.2 Image Filtering

Multi-angle measurements reduce the error caused by non-ideal characteristics of polarizers and other optical devices. However, the noise in image $I(\theta, 0)$ caused by sensitive components and external environment may not be completely eliminated using (2.61)–(2.63). From the equations for the DoLP and Orient, small noise in image $I(\theta, 0)$ will be transferred to the Stokes parameter images using (2.35)–(2.37). The noise will be present in the DoLP and Orient images. Therefore it is necessary to filter out image $I(\theta, 0)$ before calculating the Stokes images. The relative Stokes image and polarization parameter images should be calculated to prevent the amplification of noise in $I(\theta, 0)$ and the effect on the later application of polarization image.

A pre-filtering method in the process of calculating polarization parameter image is:

(1) Use a rotating polarizer to collect image $I(\theta, 0)$
(2) Filter the collected image to remove background interference. The choice of specific image filtering algorithm is determined by many factors such as the corresponding characteristics of imaging devices and specific features of external interferences.
(3) Calculate Stokes parameter images using the images $I(\theta, 0)$ after filtering

2.5.3 Image Fusion

Calculation of Stokes parameter images, DoLP, and Orient images is a fusion process. We can calculate Stokes parameter image, DoLP, and Orient images using the frame based on image fusion to improve the performance. Image filtering can also be combined with image fusion. We have proposed a calculation method of Stokes parameter images based on the adaptive non-subsampled improvement frame, which can keep abundant detailed information and make calculation process insensitive to a tiny translation of the scene as well as rapid [9–11]. Here is the step of this method.

(1) Pre-filter the image $I(\theta, 0)$ which has been collect before.
(2) Confirm the wavelet decomposition layers. The number of layers is determined according to the rotation speed of polarizer, the stability of acquisition platform and the relative movement of scene. If the rotation speed of polarizer is high (such as the use of electrically tunable polarizer, etc.), the acquisition platform is relatively stable and the relative movement of scene is slow, the decomposition layers of image $I(\theta, 0)$ can be less; otherwise, the decomposition layers of image $I(\theta, 0)$ will be more to satisfy the high-demanding Stokes parameter image.

(3) Decomposed image $I(\theta, 0)$ using the improvement framework without down-sampling. The wavelet decomposition coefficient at position (x, y) in the kth layer and lth direction is $I(\theta, 0, x, y, k, l)$.

$$I(\theta_j, 0, k, l) = \text{DWT}\{I(\theta_j, 0)\} \tag{2.66}$$

(4) Perform (2.61)–(2.63) to get the corresponding fusion coefficient in each layer on the wavelet coefficients of the image, here N represents the polarization angles:

$$\tilde{s}_0 = (x, y, k, l) = \frac{1}{N} \sum_{j=1}^{N} I(\theta_j, 0, x, y, k, l) \tag{2.67}$$

$$\tilde{s}_1 = (x, y, k, l) = \frac{2}{N} \sum_{j=1}^{N} I(\theta_j, 0, x, y, k, l) \cos(2\theta_j) \tag{2.68}$$

$$\tilde{s}_2(x, y, k, l) = \frac{2}{N} \sum_{j=1}^{N} I(\theta_j, 0, x, y, k, l) \sin(2\theta_j) \tag{2.69}$$

Reconstruct the fusion coefficients $\tilde{s}_0(x, y, k, l), \tilde{s}_1(x, y, k, l), \tilde{s}_2(x, y, k, l)$ to obtain corresponding Stokes parameter images using the inverse discrete wavelet transforms (IDWT),

$$\tilde{s}_i = \text{IDWT}\{\tilde{s}_i(x, y, k, l)\}, \quad i = 0, 1, 2 \tag{2.70}$$

Calculate the appropriate evaluation to confirm the performance of fusion results. If the evaluation satisfies with the requirements, stop and exit. Otherwise, return to step (2). We assume that there is only a certain degree of translation caused by device vibration and scene motion among the image I_0, I_{45}, I_{90} and I_{135} while the rotation is ignored, then transfer the image I_0, I_{45}, I_{90} and I_{135} for 3–4 pixels both in horizontal and vertical direction, the result is used to calculate Stokes parameter image, linear polarization and polarization phase images by (2.35)–(2.37). From Figs. 2.14, 2.15, 2.16, 2.17 and 2.18, we can find that the edge of images obtained by fusion is clearer than those calculated directly by using (2.61)–(2.63), which indicate that the fusion algorithm have better performance in preserving image details and tolerating any small translation.

In order to analyze the results in Figs. 2.14, 2.15, 2.16, 2.17 and 2.18 both qualitatively and quantitatively, we compare the difference of Stokes parameter image, the DoLP and Orient images between the two methods above, whose result is shown in Table 2.4 using four parameters: average, variance, entropy and boundary energy. Comparison can be seen in Table 2.4 that the values of entropy and boundary energy in the images obtained by fusion are higher than others.

(a)   (b)

Fig. 2.14 Parameter image s_0, **a** obtained by fusion and **b** obtained by directly calculation

(a)   (b)

Fig. 2.15 Parameter image s_1, **a** obtained by fusion and **b** obtained by directly calculation

(a)   (b)

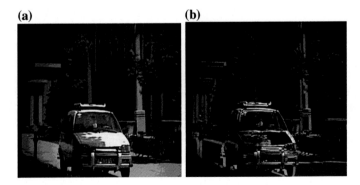

Fig. 2.16 Parameter image s_2, **a** obtained by fusion and **b** obtained by directly calculation

The former shows that fusion increases the amount of information and the obvious
change of latter proves that images which are obtained by fusion have more salient
edge information and better effect. Meanwhile, the fusion algorithm also achieves

Fig. 2.17 Parameter image *DoLP*, **a** obtained by fusion and **b** obtained by directly calculation

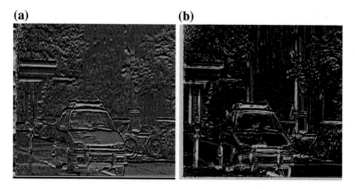

Fig. 2.18 Parameter image *Orient*, **a** obtained by fusion and **b** obtained by directly calculation

Table 2.4 Statistical comparisons of translating both horizontally and vertically

	Average	Variance	Entropy	Average boundary energy	Time (sec)
Direct calculation	101.0211	87.7710	4.9826	29.6	0.0042
Image fusion	97.6721	80.6574	5.7120	40.1	0.0142

better visual performance. Therefore, the fusion algorithm is proved to provide great convenience on calculating high-quality Stokes parameter, the degree of linear polarization and polarization angle images.

2.6 Micro-polarizer Array Imaging

Polarization imaging devices described in the previous sections require multi-angle polarization by mechanical rotation or electronic tuning. However, loss of geometric features on a moving platform is serious. Addition of polarizers to imaging sensor reduces the energy to reach the focal plane so that obtained images demonstrate serious noise interference. Comparing with the spectral characteristics, the polarization characteristics of the surface features are not obvious. High sensitivity of polarization detectors is required to take advantage of polarization characteristics effectively. These factors also bring difficulties in the application of polarization imaging detection. On the other hand, they also contribute to the emergence of new and more efficient polarization imaging devices to improve the lack of polarization imaging devices and make a wide use of polarization imaging technology.

In recent years, more practical strategies have been developed, including division of focal plane (DoFP) devices [1]. DoFP devices, also called microgrid polarimeters, provide several advantages over traditional polarimeters. Microgrid polarimeters capture images by incorporating a pixel-wise micro-polarizer array (MPA) aligned to a focal plane array (FPA), as shown in Fig. 2.19. Such imagers are akin to the Bayer pattern of color imaging sensors, except that the pattern has different polarization states, not the primary colors. This allows an MPA image to contain a full set of polarization intensity measurements to reconstruct the Stokes vector at each region of the image. Acquired images in this way are called polarization mosaic images. As additional advantages, microgrids are rugged and compact, making them ideal for real-time video imaging applications.

2.6.1 Micro-polarizer Array

Both the design and performance characteristics of the MPA are essentially determined by the calculation method of the Stokes vector calculation ways and the

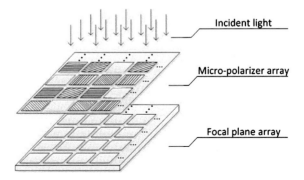

Fig. 2.19 Configuration of DoFP

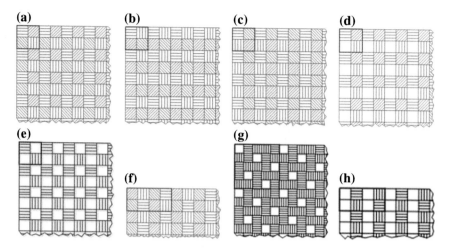

Fig. 2.20 Various MPA models: **a** Model 1 [13], **b** Model 2 [14], **c** Model 3 [12], **d** Model 4 [15], **e** Model 5 [15], **f** Model 6, **g–h** Model 7 and Model 8 [16]

arrangements of the polarization filters in the MPA. These two basic MPA features specify the construction requirements and cost.

(A) Basic Theory of MPA

Stokes vector has been efficient in the representation of polarization states of a light. Most polarization imaging sensors are designed to acquire the Stokes vector images indirectly. As shown in Fig. 2.20a–c, four micro-polarizers of 0°, 45°, 90° and 135° are placed in a repeated 2 × 2 sub-array in three MPA models [12–14]. And the fourth MPA model in Fig. 2.20d consists of three micro-polarizers and one full pass cell placed in a repeated 2 × 2 sub-array [15]. With one full pass cell added, this model shows high sensitivity of light with small polarization artifacts. Gruev et al. [12] has given the formula to eliminate Stokes parameters using $I(0), I(45)$ and $I(t)$ according to the principles of optical polarization:

$$S_0 = I(t) \tag{2.71}$$

$$S_1 = 2I(0) - I(t) \tag{2.72}$$

$$S_2 = 2I(45) - I(t) \tag{2.73}$$

where $I(t)$ denotes measured intensity through isotropic filter that passes through all the states equally. Based on (2.71)–(2.73), MPA model 5 is obtained with two micro-polarizers and two full pass filter placed in a repeated 2 × 2 sub-array, which is shown in Fig. 2.20e. With two full pass filters added, this model has higher light sensitivity and more polarization artifacts. Besides the MPA models, a repeated 2 × 4 sub-array is designed to provide better special resolution. While the MPA models designed by authors in [16] is shown in Fig. 2.20g, h.

(B) Matrix Representation of MPA in the Spatial Domain

Let $h_{MPA}(x, y)$ represent the output of an MPA sensor, sampled on a rectangular sampling lattice D. The horizontal and vertical sample spacing are equal, and this sample spacing is used as the unit length, called the pixel height. Assume that there exist underlying $I(0), I(45), I(90)$ and $I(135)$ component signals that need to be estimated. The MPA signal is obtained by subsampling and adding the four subsampled signals. The subsampling can be represented as a multiplication by subsampling functions that take the value 1 on and zero otherwise. A MPA $h_{MPA}(x, y)$ is usually a periodic tiling of a much smaller array (a red region in Fig. 2.20), called the MPA pattern $h_p(x, y)$. Using the model 1 in Fig. 2.20, a MPA pattern can be decomposed into four primary MPA patterns $h_p^d(x, y)$, each accounting for one polarization direction d. Then symbolically we can write:

$$h_p = \sum_D h_p^d(D) \tag{2.74}$$

To ensure the same dynamic range of the sensed image at all pixels, the sum of all primary MPA patterns should be a matrix with all 1's:

$$\sum_D h_p^d(x, y) = 1 \quad \forall x, y \tag{2.75}$$

For the pattern $[I(0)I(45); I(135)I(90)]$ in Fig. 2.19a can be represented as:

$$h_p^0 = \begin{bmatrix} 1 & 0 \\ 0 & 0 \end{bmatrix}, \quad h_p^{45} = \begin{bmatrix} 0 & 1 \\ 0 & 0 \end{bmatrix},$$
$$h_p^{90} = \begin{bmatrix} 0 & 0 \\ 0 & 1 \end{bmatrix}, \quad h_p^{135} = \begin{bmatrix} 0 & 0 \\ 1 & 0 \end{bmatrix} \tag{2.76}$$

(C) Frequency Structure Matrix of MPA

Inspired by the observed patterns of the spectra of MPA filtered images, we propose to represent the spectra by faithfully recording the frequency components: their frequency details and their positions. Such information can naturally be arranged in a matrix form. Inspired by the color filter array's frequency structure matrix [17], the polarization frequency structure matrix is given. Accordingly, the MPA filtered image $f(x, y)$ is composed of $f^\theta(x, y)$ images, where $\theta \in \{0°, 45°, 90°, 135°\}$. Then it can be represented as:

$$f(x, y) = \sum_\theta f^\theta(x, y) h_{MPA}^\theta(x, y) \tag{2.77}$$

where h_{MPA}^{θ} is the corresponding MPA of direction θ defined as the periodic replica of primary MPA pattern $h_p^{\theta} : h_{\text{MPA}}^{\theta}(x, y) = h_p^{\theta}$ (x *mod* n_x, y *mod* n_y). With straightforward deduction, the DFT of $f(x, y)$ can be computed by:

$$F_{MPA}(\omega_x, \omega_y) = \sum_{k_x=0}^{n_x-1} \sum_{k_y=0}^{n_y-1} \left\{ \sum_{\theta} H_p^{\theta} \left(\frac{n_x}{k_x}, \frac{n_y}{k_y} \right) \cdot F^{\theta} \left(\omega_x - \frac{n_x}{k_x}, \omega_y - \frac{n_y}{k_y} \right) \right\} \quad (2.78)$$

where $H_p^{\theta}(\omega_x, \omega_y) = \text{DFT}(h_p^{\theta}(x,y))$, $F^{\theta}(\omega_x, \omega_y) = \text{DFT}(f^{\theta}(x,y))$ and $F^{\theta}(\omega_x - \frac{n_x}{k_x}, \omega_y - \frac{n_y}{k_y})$ has been circularly shifted. Equation (2.76) implies that in the frequency domain the spectrum F_{MPA} is a multiplexing of $n_x n_y$ frequency components centered at $\left(\frac{n_x}{k_x}, \frac{n_y}{k_y} \right)$, and each component is the sum of the original spectra F^{θ} weighted by the value of the spectrum of the MPA pattern at the corresponding frequency $H_p^{\theta} \left(\frac{n_x}{k_x}, \frac{n_y}{k_y} \right)$. Based on definition of symbolic DFT, the one period matrix representation in the spatial domain can be represented as:

$$\text{DFT} \begin{bmatrix} I_0 & I_{45} \\ I_{135} & I_{90} \end{bmatrix}$$
$$= \frac{1}{4} \begin{bmatrix} I_0 + I_{45} + I_{90} + I_{135} & I_0 - I_{45} - I_{90} + I_{135} \\ I_0 + I_{45} - I_{90} - I_{135} & I_0 - I_{45} + I_{90} - I_{135} \end{bmatrix} = \begin{bmatrix} Fs_0' & Fs_1' \\ Fs_2' & Fs_3' \end{bmatrix} \quad (2.79)$$

Similarly with the frequency spectrum of a CFA-filtered image, the baseband $(I_0 + I_{45} + I_{90} + I_{135})/4$ in MPA model 1 is called polarization intensity, and other three are named polarization chroma. The frequency spectrum of the parameter in (2.79) is shown in Fig. 2.21.

With a frequency structure, we can easily find a linear relationship between the polarization intensity and polarization chrominance components and the images of the primary polarization images as:

$$\begin{bmatrix} FS_0 \\ FS_1' \\ FS_2' \\ FS_3' \end{bmatrix} = \begin{bmatrix} 0.25 & 0.25 & 0.25 & 0.25 \\ 0.25 & -0.25 & 0.25 & -0.25 \\ 0.25 & 0.25 & -0.25 & -0.25 \\ 0.25 & -0.25 & 0.25 & -0.25 \end{bmatrix} \begin{bmatrix} I_0 \\ I_{45} \\ I_{90} \\ I_{135} \end{bmatrix} = T \begin{bmatrix} I_0 \\ I_{45} \\ I_{90} \\ I_{135} \end{bmatrix} \quad (2.80)$$

where T is called the multiplexing matrix, which gives the specific composition of each component multiplexed from the image spectra of the primary polarization images, I_0, I_{45}, I_{90}, and I_{135}. The demosaicking matrix can be found from the inverse or pseudo-inverse of the multiplexing matrix: $D = T^{-1}$ if T is invertible, or $D = T^+$ if T is not square. Then the polarization images can be obtained from the multiplexed components for demosaicking in (2.80):

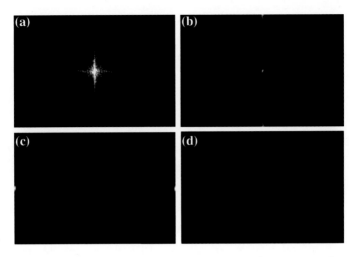

Fig. 2.21 Magnitude spectra of parameter in (2.77): **a** Parameter s_0'. **b** Parameter s_1'. **c** Parameter s_2'. **d** Parameter s_3'

$$\begin{bmatrix} I_0 \\ I_{45} \\ I_{90} \\ I_{135} \end{bmatrix} = T^{-1} \begin{bmatrix} FS_0 \\ FS_1' \\ FS_2' \\ FS_3' \end{bmatrix} = D \begin{bmatrix} FS_0 \\ FS_1' \\ FS_2' \\ FS_3' \end{bmatrix} \tag{2.81}$$

Frequency structure offers a simple universal frequency domain demosaicking algorithm for all rectangular MPAs. Aiming at minimizing the demosaicking error, some desired characteristics of an optimal MPA can be obtained, which are the principles that guide our MPA design. With the help of frequency structure, we can easily follow the design principles and turn the MPA design into an optimization problem.

(D) MPA Design Guidelines

Observation of the frequency structures matrix and general image spectra shows that:

(1) The fewer polarization components, the less frequency aliasing;
(2) The farther the distances between polarization intensity and polarization chroma and between polarization chroma, the less frequency aliasing, the polarization chroma should not be located in the horizontal or vertical axes of the intensity;
(3) Another factor that influences the accuracy of demosaicked images of the primary polarization angle is from the inverse transform from multiplexed components to the primary polarization angle: the less the norm of demosaicking matrix D, the better the demosaicking performance.

Besides the three requirements, for polarization infrared imaging system, some special requirement should be added:

(1) Cost-effective image reconstruction;
(2) Immunity to polarization artifacts;
(3) High sensitivity to light

Based on the requirements listed above, we choose a good frequency structure for less aliasing. Especially, we determine the locations of all the polarization chroma for less aliasing among polarization intensity and polarization chroma and specify some relations between the chroma, some of which may be conjugate to each other. Balancing the design expectations, we choose a new 5×5 frequency structure and a 2×4 frequency structure for our MPA design, as shown in (2.82) and (2.83):

$$S_{NewMPA1} = \begin{bmatrix} Fs_0 & 0 & 0 & 0 & 0 \\ 0 & 0 & Fs_1 & 0 & 0 \\ 0 & 0 & 0 & 0 & Fs_2 \\ 0 & Fs_2 & 0 & 0 & 0 \\ 0 & 0 & 0 & Fs_1 & 0 \end{bmatrix} \qquad (2.82)$$

$$S_{NewMPA2} = \begin{bmatrix} Fs_0 & 0 & Fs_1 & 0 \\ 0 & Fs_2 & 0 & Fs_2 \end{bmatrix} \qquad (2.83)$$

According to the above frequency structures, we propose two new MPAs, which denoted as Model 7 and Model 8, shown in Fig. 2.20g, h. To test the performance of designed MPA for small target detection, mosaic and demosaicked images of MPA Models 7 and 8. The small target is visible in the demosaicked images, as shown in Fig. 2.22.

2.6.2   Experiment Results

To determine the performance of the MPAs listed in Fig. 2.20, a number of test images have been utilized. The images used in the experiments are 320×240 images of a man on chair (Fig. 2.23a), cars (Fig. 2.23b), trees and a building (Fig. 2.23c), and water and a building (Fig. 2.23d). The test polarization images, which vary in both the complexity of the structural content (edges, fine details) and the polarization appearance, have been captured using polarization filter wheel and FLIR infrared imaging sensor. Following the evaluation procedure depicted in Fig. 2.24, tests were performed by sampling the original images with each of the MPAs shown in Fig. 2.20 to obtain a MPA image.

To evaluate the structural content performance of the considered MPAs (Fig. 2.20), image quality was measured by comparing the original infrared polarization image to the demosaicked image. To facilitate the objective

(a)   (b)   (c)

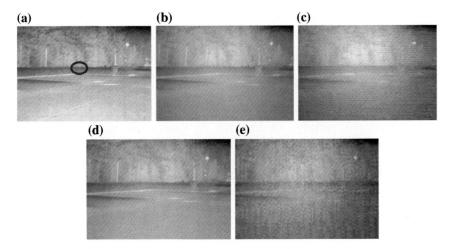

(d)   (e)

Fig. 2.22 Small target detection by using MPA model 7 and 8. **a** Infrared Intensity Image, **b** Mosaic image by model 7, **c** S_0 image calculated by using (**b**), **d** Mosaic image by model 8, **e** S_0 image calculated by using (**d**)

(a)   (b)   (c)   (d)

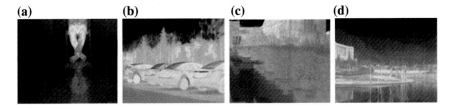

Fig. 2.23 Test images: **a** a man and chair, **b** cars, **c** trees and a building, **d** water and a building

Original image   Polarization mosaic image

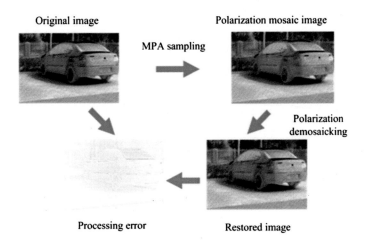

MPA sampling

Polarization demosaicking

Processing error   Restored image

Fig. 2.24 The evaluation procedure

comparisons, the polarization space based peak-signal-to-noise-ratio (PSNR) and polarization structural similarity (PSSIM) index, and the polarization feature-similarity (PFSIM) index criterion are used. Polarization PSNR can be used to evaluate the demosaicking quality. Structural information is to show that pixels with strong inter-dependencies are spatially close. These dependencies carry important information about the structure of the objects in the scene.

$$PSNR_p = 20 \log_{10}\left(\frac{S_{peak,p}}{\sqrt{MSE_p}}\right) \qquad (2.84)$$

where $S_{peak,p}$ is the average peak signal value at all polarization images. MSE_p is the average mean square error between the ground truth and the demosaicked images:

$$MSE_p = \frac{1}{MN}\|IG - ID\|^2 \qquad (2.85)$$

where M and N represent the number of rows and columns in the super-resolved image, ID is the estimated demosaicked images and IG is the ground truth polarization images.

By considering image degradation as perceived changes in structural information, PSSIM is proposed to measure the similarity between the ground truth and the demosaicked images. The PSSIM metric is calculated on various windows of an image. The measure between ID and IG of a common size $N \times N$ is [18]:

$$SSIM(IG, ID) = \frac{(2\mu_G^T\mu_D + C_1)(2\sigma_{GD} + C_2)}{(\mu_G^T\mu_G + \mu_D^T\mu_D + C_1)(\sigma_G^T\sigma_G + \sigma_D^T\sigma_D + C_2)} \qquad (2.86)$$

where μ_G is the average of IG, μ_D is the average of ID, σ_G is the variance of IG, σ_D the variance of ID, σ_{GD} the covariance of IG and ID, $C_1 = (k_1L)^2$ and $C_2 = (k_2L)^2$ are two variables to stabilize the division with weak denominator, L is the dynamic range of the pixel-values, $k_1 = 0.01$ and $k_2 = 0.03$ by default. The PFSIM metric between ID and IG is defined as [19]:

$$FSIM(IG, ID) = \frac{\sum_{z\in\Omega} S_L(z) \cdot PC_m(z)}{\sum_{z\in\Omega} PC_m(z)} \qquad (2.87)$$

where Ω means the whole image spatial domain. $PC_m(z) = \max\{PC_G(z), PC_D(z)\}$, where $PC_G(z)$ is phase congruency for a given position z of image IG. $S_L(z)$ is the gradient magnitude for a given position z.

Demosaicking results are shown in Tables 2.5, 2.6, and 2.7. Visual inspection of the demosaicked images reveals that the performance highly depends on the orientation and size of the edges and fine details in the area under consideration, and that the choice of the MPA plays an important role in obtaining the required visual quality.

Table 2.5 PSNR evaluation of images in Fig. 2.22

	Model 1	Model 2	Model 3	Model 4	Model 5	Model 6	Model 7	Model 8
Image a	36.7775	36.2677	36.3900	34.8273	34.0855	27.4510	**43.2773**	**43.1621**
Image b	33.1984	34.0372	33.8113	32.3503	31.7563	30.7389	**35.7789**	**35.3486**
Image c	33.7686	35.0067	35.2823	34.4840	34.2392	32.0621	**41.6126**	41.2439
Image d	35.1157	34.5188	34.2403	32.5251	32.0481	29.4623	**38.6929**	37.7799

Table 2.6 SSIM evaluation of images in Fig. 2.22

	Model 1	Model 2	Model 3	Model 4	Model 5	Model 6	Model 7	Model 8
Image a	0.9292	0.9237	0.9254	0.8652	0.8317	0.8877	**0.9765**	**0.9744**
Image b	0.8903	0.9056	0.9004	0.8438	0.8079	0.8497	**0.9428**	**0.9344**
Image c	0.8384	0.8987	0.8996	0.8645	0.8487	0.7404	**0.9622**	**0.9582**
Image d	0.8984	0.8952	0.8867	0.8093	0.7861	0.8797	**0.9428**	**0.9295**

Table 2.7 FSIM evaluation of images in Fig. 2.22

	Model 1	Model 2	Model 3	Model 4	Model 5	Model 6	Model 7	Model 8
Image a	0.9388	0.9324	0.9337	0.9065	0.8874	0.9021	**0.9782**	**0.9746**
Image b	0.9306	0.9383	0.9362	0.9096	0.8934	0.9144	**0.9628**	**0.9566**
Image c	0.9162	0.9402	0.9381	0.9224	0.9155	0.9743	**0.9743**	**0.9700**
Image d	0.9338	0.9309	0.9282	0.8916	0.8797	0.9191	**0.9627**	**0.9523**

Table 2.8 DoLP difference evaluation of images in Fig. 2.22

	Model 1	Model 2	Model 3	Model 4	Model 5	Model 6	Model 7	Model 8
Image a	−4.4514	−11.2548	−7.7196	12.3060	208.9453	−9.8222	**−0.6995**	**−0.7486**
Image b	−2.2497	−0.0522	−3.1469	12.9115	225.7907	−2.9522	**−0.0218**	**−0.7429**
Image c	1.3215	9.8256	−0.5207	8.4034	229.2061	−2.3474	**−0.3060**	**−0.0652**
Image d	−0.1747	−7.0869	−3.8240	7.2815	221.0587	−5.6740	**0.0797**	**0.0043**

Comparing the results in Tables 2.5, 2.6, 2.7, 2.8 and 2.9, the proposed Model 7 and 8 showed excellent results in terms of traditional evaluation criteria such as PSNR, SSIM, and FSIM, which are also effective on small targets as well as DOLP and Orient similarity.

Table 2.9 Orient difference evaluation of images in Fig. 2.22

	Model 1	Model 2	Model 3	Model 4	Model 5	Model 6	Model 7	Model 8
Image a	3.8225	19.9316	−19.0552	8.7935	53.9593	8.5967	**2.3024**	**1.9196**
Image b	2.7393	27.4827	−23.5338	31.9140	49.7365	−2.9205	**0.8424**	**1.6337**
Image c	8.3703	41.3142	−30.0654	25.6698	41.6397	−4.4156	**−7.4222**	**−4.9806**
Image d	5.2951	14.1395	−26.3736	3.8146	43.5596	−1.3936	**−1.1737**	**0.0483**

References

1. Born M, Wolf E (1999) Principles of optics, 7th edn. Cambridge University Press, Cambridge
2. Liao Y (2003) Polarization optics. Science Press, Beijing
3. Zhao Y, Gong P, Pan Q (2008) Object detection by spectropolarimeteric imagery fusion. IEEE Trans Geosci Remote Sens 46(10):3337–3345
4. Zhao Y, Zhang L, Zhang D, Pan Q (2009) Object separation by polarimetric and spectral imagery fusion. Comput Vis Image Underst 113(8):855–866
5. Zhao Y, Pan Q, Chen Y, Zhang H (2005) Clutter reduction based on polarization imaging technology and image fusion theory. Chin J Electron 33(3):433–435
6. Filippidis A, Jain LC, Lozo P (1999) Degree of familiarity ART2 in knowledge-based landmine detection. IEEE Trans Neural Networks 10(1):186–193
7. Katkovaky LV, Belyaev BI, Belaev YV, Kurikina TM, Nesterovich AI, Sosenko VA, Sosenko VA (1999) Spectropolarizational technique for detection of manmade objects in visible and near infrared spectral ranges. Proc IEEE Int Geosci Remote Sens Symp 2(8):1381–1383
8. Cremer F, Schwering PBW, De Jong W, Schutte K (2001) Infrared polarisation measurements of targets and backgrounds in a marine environment. Proc SPIE 4370:169–180
9. Zhao Y, Pan Q, Zhang H (2006) New polarization imaging method based on spatially adaptive wavelet image fusion. Opt Eng 45(12):123202
10. Zhao Y, Pan W, Zhang H (2005) Adaptive polarization image fusion based on regional energy dynamic weighted average. Optoelectron Lett 1(3):224–227
11. Zhao Y, Pan Q, Zhang H (2005) Calculation method of stokes images based on sequence image fusion. J Optoelectron Laser 16(3):354–357
12. Gruev V, Perkins R, York T (2010) CCD polarization imaging sensor with aluminum nanowire optical filters. Opt Express 18(18):19087–19094
13. Gao SK, Gruev V (2011) Bilinear and bicubic interpolation methods for division of focal plane polarimeters. Opt Express 19(27):26161–26173
14. Wu Z, Powers PE, Sarangan AM, Zhan Q (2008) Optical characterization of wiregrid micropolarizers designed for infrared imaging polarimetry. Opt Lett 33(15):1653–1655
15. LeMaster DA, Hirakawa K (2014) Improved microgrid arrangement for integrated imaging polarimeters. Opt Lett 39(7):1811–1814
16. Zhao Y, Liu W, Kong SG, Liu Y, Pan Q (2015) Design and performance analysis of infrared micro-polarizer array. In: Proceedings of the 34th Chinese control conference, July 2015
17. Dubois E (2005) Frequency-domain methods for demosaicking of Bayer-sampled color images. IEEE Signal Process Lett 12(12):847
18. Wang Z, Bovik AC, Rahim Sheikh H (2004) Image quality assessment: from error visibility to structural similarity. IEEE Trans Image Process 13(4):600–612
19. Zhang L, Mou X (2011) FSIM: a feature similarity index for image quality assessment. IEEE Trans Image Process 20(8):2378–2386
20. Azzam RMA (1985) Arrangement of four photodetectors for measuring the state of polarization of light. Opt Lett 10(7):309–311

Chapter 3
Multi-band Polarization Imaging

Establishing a multi-band polarization imaging system is the premise of acquiring accurate multi-band polarization information. However, the acquired polarization and spectral data can be noisy and less accurate in the sensing process due to the factors such as dark current, positioning error, and the transmittance in different wavelengths [2, 3]. The polarization calibration and inversion methods are also presented in detail to establish an accurate relationship between the measured data and actual polarization and spectral information, which is useful to acquire accurate polarization and spectral characteristics in multi-band polarization imaging systems [1].

3.1 Multi-band Polarization Imaging Methods

3.1.1 Spectral Band Tuning

Spectral components can be extracted using optical filters or spectral separating elements including filter wheels, tunable filters, diffraction grating, and prism. A programmable filtering element is preferred in spectral imaging of fast-moving objects in the scene. This section discusses popular spectral tuning methods such as prism and grating dispersion, interference Fourier transform, electronically tunable filters, and tomography.

(A) Dispersion with Prism and Grating

Imaging spectrometers based on dispersion with prism and grating have been widely used in aviation and aerospace fields. Spectral information of an incident light can be extracted using prism and grating dispersion. Figure 3.1 describes how prism and grating dispersion works in multi-band spectral imaging. Dispersion occurs on a prism or blazed grating after an incident light comes from the slit directly or through collimator lens. A virtual spectral image of the target can be

© National Defense Industry Press, Beijing and Springer-Verlag Berlin Heidelberg 2016
Y. Zhao et al., *Multi-band Polarization Imaging and Applications*,
Advances in Computer Vision and Pattern Recognition,
DOI 10.1007/978-3-662-49373-1_3

Fig. 3.1 Schematics of
a prism collimation imaging
system and **b** grating
dispersion imaging systems to
obtain spectral information

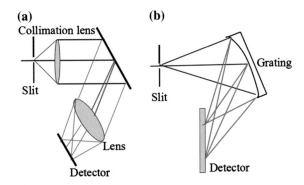

obtained in different positions on plane array detectors according to the wavelength
of the incident light [1, 3]. As shown in Fig. 3.1a, an incident light passed through
collimation lens is scattered by the prism, and the detector receives the captured
image. In Fig. 3.1b, collimation lens is not necessary in grating dispersion, where
the grating can receive the incident light directly. And the virtual image of target
will be shown on the detector with the help of grating diffraction.

(B) Interference Fourier Transform

Utilizing the Fourier transformation of spectral pixel interferogram and spectral
image, Fourier spectrometers can obtain spectral information of an object by
computing interference images and their Fourier transforms. There exist three
methods to obtain spectral pixel interferograms: Michelson interferometry, Triangle
common path interferometry, and Birefringence interferometry. Fundamental idea
of these methods is to divide incident light to form a coherent light source. Since a
monochromatic light in different wavelengths of the same optical path length can
produce different phase, the formation of interference fringes can be different. And
maximum light intensity caused by the interference in different wavelengths have
different positions in the plane array. Triangle common path interferometer and
birefringence interference imaging spectrometers belong to a type of spatial modulation. Both interferometers can be used for spatial- resolved spectrum and
time-resolved spectrum, which are suitable for spectral earth remote-sensing
without registration. Generally, spectral information needs to be measured at least
in three polarization directions. Effective methods and equipment are not yet
available to combine Fourier transform spectral detection and polarization
detection.

(C) Electronically Tunable Filters

Optical filters is the most simple and popular instrument in band selection. A set of
spectral images in different wave bands can be obtained in the imaging plane by
changing the center frequency of optical filters. Often optical filters are installed on
a mechanical filter wheel to make an imaging process convenient. A stepper motor
rotates the wheel to align a filter to the light path. Such a mechanical way of tuning

involves an inevitable delay in response as well as vibration and displacement that need to be dealt with in the imaging process.

Various types of tunable filters such as acousto-optic tunable filters and liquid crystal tunable filters have been used to separate spectral bands in multi-band polarization imaging systems. An AOTF is composed of the elements of good optical performance, low audible sound and light attenuation according to the sound and light diffraction theory. A variable diffraction grating is created by applying radio frequency waves to a piece of birefringent crystal which determine the type of AOTF as linear or non-linear. In the structure of linear AOTF, incident light, diffracted light and acoustic wave are carried out in the same direction, while the direction will be different in non-linear ones. A diffraction beam drifting upward and downward is orthogonally polarized and separated from non-diffracted beam in space when emitted from crystal. Adjusting the frequency of the sound wave completes a spectral scanning in spectral range, so the polarization characteristics of target can be measured by placing a phase delay plate in front of AOTF. A Lyot-type tunable birefringent LCTF is sequentially concatenated by multiple stages arranged in parallel. Tuning electrical filter can change the wave band in spite of the cost of switching time and imaging efficiency, which makes the entire imaging system simple and compact. The effect of switching time can be reduced in still imaging [2].

(D) Tomography

Tomography imaging spectrometers have attracted interests in the field of spectral imaging. It captures spectral images in a form of three-dimensional data cube, and a special spectrometer is used to record the projection of image cubes, which will be analyzed to reconstruct the spatial and spectral images [1]. One of the most significant advantages of tomography imaging spectral technology is the full field, which not only makes a full use of light source, but also provides image cubes in hyperspectral resolution. Moreover, tomography imaging also has a short detection time, high-flux and high stability. Tomography imaging spectrometers can be divided into grating based spectrometers and prism based spectrometers according to different function modes. Imaging spectrometers with high-flux grating can obtain images of a 2D target with spatial and spectral information. While prism based spectrometers obtain projection images of the target in multiple directions, which can be used to reconstruct spatial and spectral images by tomography processing [1, 3].

3.1.2 Polarization Adjustment

An easiest way to change polarization direction is to rotate an optical polarizer mounted in front of object lens. Birefringent mica polarizer is compact and light with good optical performance, which can obtain three equations at three polarization directions to solve polarization characteristics. To obtain images of the same

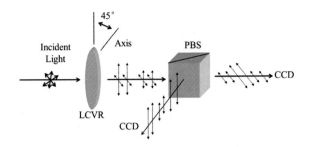

Fig. 3.2 Measurement of parameter s_1 by PBS and LCVR

scene in different polarization angles, an optical spectroscopic method separates incident light into three beams that go through polarizers of different directions. Comparing with the case of multi-camera, optical spectroscopic method has higher system performance requirement but with no need of image registration. LCVR has better effect by changing the range of liquid crystal array in the process of wavelength changing. A polarizing beam splitter (PBS) prism can separate non-polarized incident light, which means that P-polarized light goes through while S-polarized light is reflected. PBS is designed using *MAC Neille* method mentioned in [10], which makes each layer surface to satisfy the Brewster angle so that the transmittance of P-polarized light becomes nearly 100 %. As for S-polarized light, the layer surface constitutes a highly reflected film system that achieves the theoretical reflectivity of 100 %.

We coordinate LCVR with a splitter prism, so polarization can be detected without any polarizer. Taking the measurement of Stokes parameter s_1 as an example, an LCVR is placed in front of splitter prism, the optical axis and the horizontal direction are at the angle of 45°, as shown in Fig. 3.2. LCVR converts zero-wave to half-wave. Polarization of an incident light will not change at zero-wave while vertical component will be rotated to a parallel one at half-wave. So the Stokes parameter s_1 can be measured. Usually the formula of polarization parameters is very complex and sensitive to noise in the way of changing the amount of delay. So in the actual polarization detection system, a polarizer is usually used for measuring polarization to simplify the calculation of polarization parameters and eliminate a part of systematic errors.

3.2 Multi-band Polarization Imagers Based on Tunable Filters

Current mainstream multi-band polarization imaging systems change the spectral band using tunable filters which have fast response and simple structures. Combining tunable filters with polarizers or retarders is relatively easy. Commonly used tunable filters include LCTFs and AOTFs, which can achieve spectral scanning and measure in coordinate with LCVR.

3.2.1 Liquid Crystal Tunable Filters

A Lyot-type tunable birefringent LCTF is sequentially concatenated by a multiple stages arranged in parallel. Each level includes two parallel polarizers which have a liquid retard between them. The axis of retarder and the pass direction of polarizer are at the angle of 45°. The thickness of retarder increases progressively, which is given by d, $2d$, ..., $2(N-1)d$, where d denotes the thinnest one. Incident polarizer at the first level divides the incident light into two equal components in different crystal axes where the relative retardation will lead to phase difference. Voltage applied to the liquid crystal elements changes refractive index at one of crystal axes. These two components can be reassembled into a predictable polarization again on the exit polarizer. Transmittance of each level is a function of wavelength.

$$I(\lambda) = I_0 \cos^2 \frac{\Gamma(\lambda)}{2} \tag{3.1}$$

$\Gamma(\lambda)$ is a function of incident angle θ and azimuth ϕ, representing the retardation at the level of thickness d. The total transmittance of a multi-level is like 2^N lines raster which generates factor $\cos^2(\cdot)$. The transmittance of an N-level LCTF is

$$T_f = \prod_{i=1}^{N} \cos^2\left\{2^{i-1}\Gamma(\lambda)/2\right\} = \frac{\sin^2 2^N \Gamma(\lambda)/2}{2^{2N} \sin^2 \Gamma(\lambda)/2} \tag{3.2}$$

The liquid crystal in Fig. 3.3 achieves wavelength tuning. In normal incidence conditions, total transmittance of one level can be represented by

$$\Gamma = \Gamma_0 + \Gamma_1 + \Gamma_2 \tag{3.3}$$

Fig. 3.3 A schematic diagram of a Lyot filter

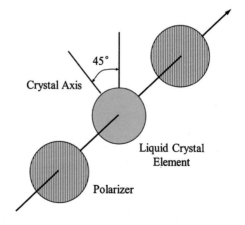

where $\Gamma_0 = 2\pi(\Delta n)d/\lambda$ denotes the retardation of a liquid crystal element which has thickness d and birefringence Δn. Γ_1 and Γ_2 represent the retardation of lithium niobate crystal in different directions. The effect of half wave plate is ignored in (3.2). To achieve tuning in the whole independent spectral region, Γ_0 should have a range changing from 0 to 2π. So the birefringence of liquid crystal is λ/d.

3.2.2   Acousto-Optic Tunable Filters

According to the principle of acousto-optic tunable filters, a non-collinear acousto-optic tunable filter has tuning relationship that

$$\lambda_0 = v_\alpha \Delta n (\sin^2 2\theta_i + \sin^4 \theta_i)^{1/2}/f_\alpha \qquad (3.4)$$

where λ_0 denotes the diffraction wavelength, v_a is ultrasonic velocity, θ_i is the angle between incident light and crystal axis, f_a is drive frequency of ultrasonic wave and Δn is the birefringence of interaction medium. Tuning relationship reflects one-to-one correspondence between driving frequency and diffraction wavelength. Spectral scanning can be completed by adjusting wave frequency electronically in the range of certain spectrum.

3.2.3   Liquid Crystal Variable Retarders

An LCVR is a phase tunable retarder with a big aperture, which is made by arranging nematic liquid crystal between the prepared substrates. An LCVR can divide a light into two beams of orthogonal polarization direction with different phases. LCVR only changes the polarization of light in an ideal condition. Retardation phase is $\delta = (\Delta n)d/\lambda$ without applying voltage, Δn is the birefringence of liquid crystal, d denotes the thickness of liquid crystal layer. Nematic liquid crystal has an electrical dipole moment. Electric field is applied to make positive dielectric anisotropy nematic liquid crystal molecules arrange parallel to the direction of electric field. Incident retardation phase is continuously adjustable by changing the direction of voltage continuously adjustable axis at the two ends of LCVR liquid crystal. The maximum of retardation phase will decrease with the increment of voltage and it can be acquired when voltage equals zero. An LVCR has a compensation function. A single LVCR satisfies phase retardation of different wavelength and calibrates in the certain wavelength to get highly accurate retardation. Its delay effect has little to do with the incident angle and the power is low voltage AC. The response time can reach the in milliseconds level. All such characteristics make the foundation of accurate and efficient polarization measurement.

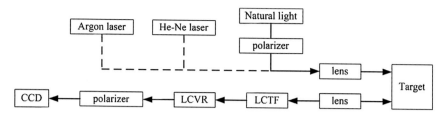

Fig. 3.4 A multi-band polarization detection system

3.2.4 Typical Multi-band Polarization Imaging Systems

Currently the detection systems applied in multi-band polarization remote sensing are either active or passive [4]. Active ones use laser source, converting polarization direction of incident light independently so that the polarization information of the corresponding target is strengthened. The source of active system is laser while the source of passive system is sunlight.

An active multi-band polarization imaging system uses standard and halogen lamps to obtain multi-band polarization information in the laboratory. Figure 3.4 shows the structure of a multi-band polarization system in Taylor laboratory, France. Argon laser and He–Ne laser are used in this system, transmitted wavelength and polarization of reflected light are adjusted by LCTF and LCVR. Combining active and passive multi-band polarization detection systems is a research topic to improve the performance of classification and detection.

Passive multi-band polarization imaging has been the mainstream. Early spaceborne multi-band polarization detection systems such as POLDER of France acquire the information by rotating filter wheel in a few wave bands [2]. The performance of imaging system has been greatly improved and still evolved when electronic tuning is applied to multi-band polarization detection. The structure of multi-band polarization imaging system studied in ARL is shown in Fig. 3.5. LCVR is used to adjust the polarization of incident light which changes to parallel beam when passes through camera lens. After phase retardation and wave bands tuning by LCVR and AOTF respectively, the light is reflected by right-angled plane mirror and imaged at CCD plane array [5, 6]. All of these multi-band polarization detection systems

Fig. 3.5 The structure of multi-band polarization imaging detection system based on AOTF

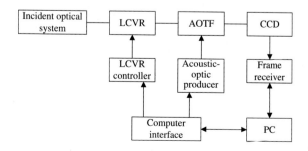

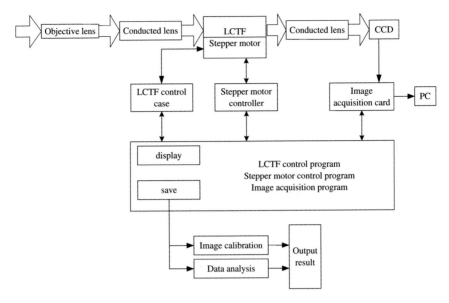

Fig. 3.6 Structure of multi-band polarization imaging detection system in National Aeronautical Laboratory of Japan

achieve great effect in actual applications. The system in Taylor laboratory works both actively as well as passively. To make the light consistent from the source, incident natural light is allowed going through a linear polarizer to be linear polarization light so that multi-band polarization performance can be studied in the direction of eliminating polarization. The notable feature of the imaging system in Figs. 3.4 and 3.5 is using the basic units of polarization detection: polarizer and phase retarder. So the component s_3 in Stokes vector can be detected and mechanical shaking can be eliminated with the help of electrically controlled adjustment. Both polarization detections are based on the phase delay of LCVR, which make the calculation formula of polarization parameter complicated [9].

National Aeronautical Laboratory of Japan has developed a multi-band polarization imaging system using LCTF. It can obtain multi-band polarization images freely in the chosen range of wavelength bands and polarization directions, which can identify the characteristics of target surface. Figure 3.6 shows the structure and imaging process of this system [7, 8].

The main component of multi-band polarization imaging detection system in National Aeronautical Laboratory of Japan is LCTF. The component s_3 in Stokes vector cannot be detected without any phase delay devices, but it can get components s_0, s_1, s_2 by rotated linear polarizer. Comparing with the first two detection systems, this system has many advantages. Firstly, the structure is easy to achieve. LCVR is not applied in the system so that optical path effect can be ignored in system establishment, so the imaging optical path will be much simplified. Moreover, the costs can also be reduced by leaving out LCVR. Secondly, the calculation of polarization parameters is easy, which only includes difference and sum operation without multiplication and division, so the error caused by outside interference is reduce. And difference operation also compensates for system errors to ensure the accuracy of polarization parameters.

3.3   Multi-band Polarization Imagers Using LCTF

3.3.1   Implementation of Multi-band Polarization Imaging Systems

A multi-band polarization imaging detection system in Fig. 3.7 consists of an LCTF with the spectral range of 400–700 nm and a 12-bit high power gain CCD controlled by the computer. The system can display, collect, and save images at selected wavelength bands at a specific polarization direction. As a result, the cycle acquires multi-band intensity images in different polarization direction during a few seconds. The imager is easy to operate and offers clear image quality and high spectral resolution. As shown in Fig. 3.7, after going through the first group of lenses, the incident light I_{total} becomes a collimated monochromatic parallel light which is accumulated and imaged on the CCD. The CCD sensor and the computer are directly connected with a FireWire port, or IEEE 1394 serial port. Since a direct connection lacks electrical power, a powered hub is necessary to ensure stable operations of the CCD. The filter is connected with USB through the control box. A high-gain camera by QIMAGING compensates for wave energy attenuation during filtering. In the inversion system, a polarization spectrometer is consists of a polarizer and a portable spectrometer connected with computer. So the radiation intensity curve and relative reflectance curve of a pixel can be directly acquired.

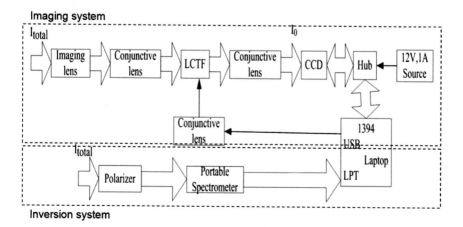

Fig. 3.7 Structure of a multi-band polarization imaging detection system

The data collected by polarization spectrometer is used for polarization inversion. Main parts of the system are installed on a PTZ platform. Because of its small size, light weight, and convenience, the system is totally suitable for data collection tasks in the field. The filter is equivalent to the combination of a polarizer and a filter with tunable transmitted wavelength range of the transmittance spectral band. I_θ is acquired by I_{total} through the filter at 0°, 45°, and 90°. The polarization direction of the filter is parallel when $\theta = 0°$. The filter is rotated clockwise [6]. The Muller matrix of the filter is denoted as **M**, which can be solved by

$$s_{out} = \begin{bmatrix} s_{out,0} \\ s_{out,1} \\ s_{out,2} \\ s_{out,3} \end{bmatrix} = \begin{bmatrix} m_{11} & m_{12} & m_{13} & m_{14} \\ m_{21} & m_{22} & m_{23} & m_{24} \\ m_{31} & m_{32} & m_{33} & m_{34} \\ m_{41} & m_{42} & m_{43} & m_{44} \end{bmatrix} \begin{bmatrix} s_{in,0} \\ s_{in,1} \\ s_{in,2} \\ s_{in,3} \end{bmatrix} = \mathbf{M} s_{in} \qquad (3.5)$$

The multi-band polarization imaging detection software is developed based on the SDK by LCTF and CCD components, which mainly contain the functions of filter port inquiry, band selection, band auto scan, CCD parameters setting, images display, images capture, and automatic storage. The software operation can be described as:

(1) Initialize the filter;
(2) Adjust the CCD parameters according to the grey value in different wave bands;
(3) Capture and save the images.

Since the transmittance varies in different wavelengths as well as polarization directions in the imaging process, the grey values show big differences among different groups of multi-band polarization images. The lower the grey value is, the lower the SNR. Too large gray values may exceed the saturation level of the CCD. Therefore it is important to set suitable exposure time and camera gain. Camera settings will be completed in this step and remain unchanged during image collection. Images will not be displayed in the acquisition and saving processes. The software has clear function for convenient operations.

3.3.2   Main Devices of Multi-band Polarization Imaging Systems

Table 3.1 shows the specifications of the multi-band polarization imaging detection system (VariSpec VIS filter, CRI). Tuning response time of the filter mainly relies on ambient temperature in actual detection. Low temperature can slow down the response speed of the filter. The width of a wavelength band is defined as the FWHM of transmittance curve. The wave width of a Lyot filter is fixed when adjusted in the range of a certain wavelength band. A high-gain, high-resolution

Table 3.1 Main parameters of VariSpec VIS filter

Wavelength range	400–1000 nm
Band width	7 or 10 nm
Operation aperture	20 mm
Field of view	7.5° (from the normal)
Band accuracy	1/8 bandwidth
Maximum in-out irradiance	500 mW/cm^2
Tuning response time	50 ms

Table 3.2 Main parameters of Qimaging CCD

Sensor type	Sony ICX205
Exposure time	12 µs–17.9 min
Resolution	1.4 million pixels
Pixel size	4.65 µm × 4.65 µm
Readout frequency	20, 10, 5, 2.5 MHz
Frame rate	10 frames/s
Data transfer rate	40 MB/s
Gain control	0.6–15 dB
Offset	−2048 to 2047
Range of object and image distance	≥61 mm

CCD can reduce the difficulties such as attenuation in filtering and unsatisfactory imaging results. An optical imaging system focuses the light reflected from the target on the photosensitive surface of the CCD, where photosensitive cells covert the light intensity into electrical charges. The video signals of the target will be acquired from the CCD output. Every discrete voltage signal in the video signals corresponds to the image intensity on photosensitive cells. The signal output order is determined by the spatial position of the CCD photosensitive cells so that CCD can convert information from spatial domain to time domain in the way of self-electronic scanning.

A multi-band polarization image collection system by QIMAGING offers a high-speed and high-resolution 12-bit digital CCD that connects to the computer by IEEE 1394. Table 3.2 summarizes the specifications of the parameters.

A portable spectrometer GRE 2000 is a convenient spectroradiometer operated using a 12 V battery. It can collect spectral data from 350 to 2500 nm. Fast scanning capability enables operations in outdoor and aviation as well as in the laboratory. Table 3.3 shows the parameter specifications.

3.3.3 Multi-band Polarization Imaging Sensors

The multi-band polarization imaging detection system shown in Fig. 3.8 is mounted on the tripod during image acquisition. We use a laptop computer in Windows XP

Table 3.3 Main parameters of GRE 2000

Spectral range	350–2500 nm
Number of channels	640
Sensor type	512Si, 128PbS
Sampling bandwidth	300–1050 nm:1.5 nm 1050–2050 nm is 11.5 nm
Size	286 mm × 305 mm × 114 mm
Field of view	10 frames/s

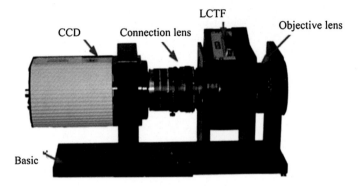

Fig. 3.8 A multi-band polarization image detection system

and the software Visual C 6.0. A repeater hub is used to connect CCD and IEEE 1394 port. Parallel beams are filtered in the imaging optical path as long as the lens and the front conducting lens have the same focal plane. Multi-band polarization images in different directions are formed at the time of filter rotation.

Figure 3.9 shows an experiment device used to acquire multi-band polarization curve of the target. The lens of spectrometer and multi-band polarization imaging detection system should be kept in the same direction to ensure the same polarization of light.

Fig. 3.9 A multi-band polarization data collection device

(a)

(b)

(c)

(d)

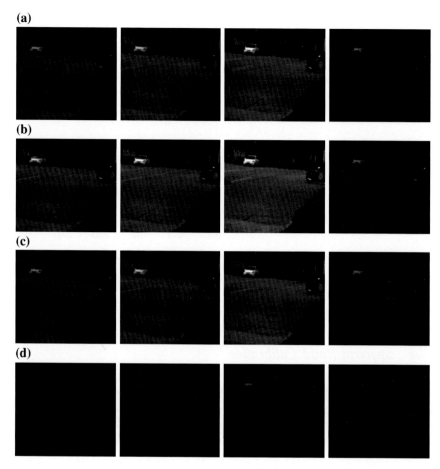

Fig. 3.10 Example multi-band polarization images in four polarization directions and at **a** 460 nm, **b** 540 nm, **c** 580 nm, **d** 640 nm, from the *left* to *right* are 0°, 45°, 90°, 135°

The multi-band polarization image detection system is used to collect a library of multi-band polarization images. The database contains more than 100 groups of experimental images of different scenes, a set of 132 multi-band polarization images from 400 to 720 nm in 33 wavebands, and a panchromatic image. Figure 3.10 shows multi-band polarization images in four polarization directions, 0°, 45°, 90°, and 135° at the wavelengths of 460, 540, 580, and 660 nm. Figure 3.11 shows the intensity of polarized reflectance light of grass as a function of the wavelength at the polarization direction of 45°.

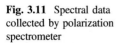

Fig. 3.11 Spectral data collected by polarization spectrometer

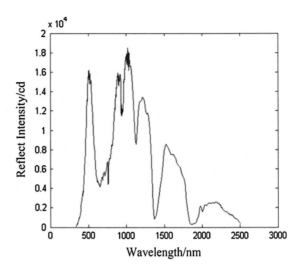

3.4 Calibration of Multi-band Polarization Imaging Systems

3.4.1 Spectral Band Calibration

The method of calibrating spectral bands is to ensure the wave width of camera spectral response curve in the condition of different tunable wavelength. Multi-band polarization remote sensing devices have multi-channels and high spectral resolution so that the calibration of wavelength is strict. The device in the range of reflecting spectral should set the low-pressure mercury lamp and krypton lamp emission line as standard. Monochromator is calibrated, after which the monochromator is made to output single light in a certain step, simultaneously the output signals are detected and recorded by sensor [1]. By comparing the output signal of the monochromator and the wavelength location, the curve shape sensor of measurement signals acquired by sensor, we can determine the wavelength location of each channel and the spectral response function. The multi-band polarization imaging detection system has high spectral resolution than other similar optical sensors.

Figure 3.12 shows the CCD output spectral radiation intensity when the emitted laser from 632.8 nm He–Ne gas laser incident to the objective lens. Intensity values have already been normalized so that maximum values are the same. FWHM is usually used to represent spectral resolution. The result shows that the spectral resolution of optical sensor is lower than 10 nm at the wavelength of 632.8 nm, moreover, it is higher in a small range of wavelength and lower in a wide range and spectral resolution of optical sensor is 5–10 nm from 400 to 700 nm.

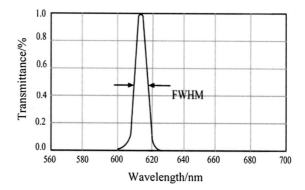

Fig. 3.12 Transmittance of a He–Ne laser

3.4.2 Spectral Radiometric Calibration

Radiometric calibration is a significant part of calibration. The AVIRIS radiometric calibration system is a complete system that has a 100 W quartz halogen installed to apply fixed current and voltage, using standard reflector to reflect light to the instrument for measuring instantaneous field of view. Radiometric calibration coefficient is calculated through the ratio of radiation intensity reflected to view and values measured by devices. Scattered light must be blocked completely so that the scattered light cannot be irradiated to the reflection board and the dichroic reflection characteristics are measured. We use a 10 m diameter, 30 cm opening integrating sphere as a light source irradiating the entire field of sensor and simultaneously measure and calibrate the homogeneity of radiation characteristic, as well as radiation accuracy and the SNR [1].

The radiation calibration method in this context is designed following the radiometric calibration method of AVIRIS. The intensity of light incident to object lens can be acquired by measuring CCD output spectral radiation in initial calibration. Therefore an integrating sphere with known spectral radiation and a white board with known reflectivity should be used. The light emitted from the integrating sphere through a whiteboard is reflected into the lens of an optical sensor. The radiation of object lens can be calculated using spectral radiation, reflectivity of white board, and the length of optical path. The corresponding CCD output spectral radiation is acquired by image data, the relationship between calculated spectral radiation value and the measured one determined the spectral radiation calibration.

3.4.3 Mechanical Revolving Filters

To ensure the LCTF rotating mechanical properties, a linearly polarized light is irradiated onto the lens to measure the amount of CCD output radiation. In a polarization performance test, the polarization direction of the linearly polarized light is horizontal. And initial polarization direction of LCTF is vertical. The CCD

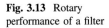

Fig. 3.13 Rotary
performance of a filter

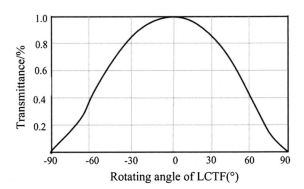

output radiation changes with the angle turned by LCTF. So the rotary performance is reliable and accurate as long as CCD output radiation is the minimum at ±90° and the maximum at 0°. Figure 3.13 shows the rotary performance curve in the rotating angle of LCTF. The vertical axis denotes the ratio of linearly polarized light radiation and multi-band polarization detection radiation.

3.4.4 *Polarization Calibration*

Detecting polarization model can predict an ideal polarization response. Polarization of an incident light is easily influenced by the noise [2]. The performance and response relationships also change with different scenes. Calibrating the entire sensors accurately can be viewed as a polarization characteristics calibration that concludes the radiation response formula, Stokes polarization response inverse matrix, and a radiation model at every pixel. A polarization calibrating source confirms the polarization at the time of calibrating and testing to detect the polarization accuracy and to monitor the change of device characteristics. So the polarization calibrating source must have polarization accuracy higher than the instrument sensitivity such as He–Ne laser (632.8 nm, less than 2 mW), a polarizer, and a 1/4 wave plate [3]. After the desired non-polarized light from the polarizer and polarization detection system, the normalized Stokes vector of emitted light is:

$$S_{PSG} = M(\beta)M_p(\theta)S_{un} \begin{bmatrix} 1 \\ \cos 2\beta \cdot \cos(2\beta - 2\theta) \\ \sin 2\beta \cdot \cos(2\beta - 2\theta) \\ \sin(2\beta - 2\theta) \end{bmatrix} \qquad (3.6)$$

where S_{un} denotes the Stokes vector of non-polarized light; $M_p(\theta)$ is the Muller matrix of polarizer with azimuth θ; $M(\beta)$ is the Muller matrix of wave plate with azimuth β. From (3.6), a polarized calibrating source can produce arbitrary complete polarization by adjusting azimuth θ and β. In an ideal condition, we use s_0, s_1, s_2 to represent polarization produced by polarized calibrating source to detect the multi-band polarization images. From (3.1) that every pixel of the detector has response at $0°$, $45°$, and $90°$.

$$I_0 = \frac{1}{2}(s_0 + s_1)$$
$$I_{90} = \frac{1}{2}(s_0 - s_1) \Rightarrow \begin{bmatrix} I_0 \\ I_{90} \\ I_{45} \end{bmatrix} = \begin{bmatrix} \frac{1}{2} & \frac{1}{2} & 0 \\ \frac{1}{2} & -\frac{1}{2} & 0 \\ \frac{1}{2} & 0 & \frac{1}{2} \end{bmatrix} \begin{bmatrix} s_0 \\ s_1 \\ s_0 \end{bmatrix} \qquad (3.7)$$
$$I_{45} = \frac{1}{2}(s_0 + s_2)$$

Solving the inverse matrix of polarization response matrix and the response to any solution with equal weight is given by

$$\begin{bmatrix} s_0' \\ s_1' \\ s_2' \end{bmatrix} = \begin{bmatrix} 1 & 1 & 0 \\ 1 & -1 & 0 \\ -1 & -1 & 2 \end{bmatrix} \begin{bmatrix} I_0 \\ I_{90} \\ I_{45} \end{bmatrix} \qquad (3.8)$$

s_0', s_1', s_2' are calculated values of an incident beam. The general formula to solve polarization parameters by Stokes polarization response inverse matrix is as follows.

$$\begin{bmatrix} s_0' \\ s_1' \\ s_2' \end{bmatrix} = \begin{bmatrix} K_{I,1} & K_{I,2} & K_{I,3} \\ K_{Q,1} & K_{Q,2} & K_{Q,3} \\ K_{U,1} & K_{U,2} & K_{U,3} \end{bmatrix} \begin{bmatrix} I_0 \\ I_{90} \\ I_{45} \end{bmatrix} \qquad (3.9)$$

where $K_{i,j}$ is determined by remote sensing calibration. The calibrated data is acquired by recording the remote sensing data at the rotation position of a few different polarizers. Stokes parameters of light source will be used in multivariate regression analysis to get Stokes polarization response matrix.

3.5   Reconstruction of Multi-band Polarization Imagery

Received signals cannot always reflect the characteristics of the surface because of the factors such as Rayleigh scattering and Mie scattering of air molecules and aerosol, the surface, the terrain, and the solar radiation spectrum effect [1, 2]. The influence caused by the air should be considered first. Polarization inversion was designed to eliminate the effects, making the surface parameters inversion of polarization practical.

3.5.1   Spectral Inversion Methods

(A)   Atmospheric Radiative Transfer Correction by Radiation Transfer Equation

Since atmospheric scattering and shape of the solar spectrum can be determined by radiative transfer model, the LOWTRAN 7 reflectance model is given by

$$R = (L - L_0)/L_x \tag{3.10}$$

where R denotes the reflectivity of a certain pixel of a certain waveband, L is the corresponding DN value. The coordinates, data acquisition time, optical thickness, and atmospheric moisture content are assumed known in this method.

(B) Reflectivity Inversion by Images

This method is mainly carried out based on image data. Typical approaches are as follows:

(1) Internal Average
     DN values of the image band are used to divide the average of this band to get the relative reflectivity, which needs to have complete scene type. The average spectral curve of entire image does not have apparent absorption. Artificial noise is easily scaled as spectral characteristics so that some of the actual spectral characteristics may be compressed or even lost. This depends on spectral characteristics of the corresponding parts which play a major role in imaging. Relative value of reflectivity can be acquired using this method.
(2) Flat-field Domain
     To find an area with high brightness and smooth spectral response curve, DN value is used to divide the average spectral response to reduce atmospheric effects, some main characteristics of the gas absorption, and the residual effects of the instrument. This method has some limitations because the ground targets like pond and cement pavement, and their multi-band polarization characteristics should be known.
(3) Logarithmic Residuals
     Logarithmic residuals can eliminate the effect of sunlight, atmospheric and terrain impact. The radiation DN_{ij}, reflectivity R_{ij}, terrain factor T_i, and illumination factor I_j at pixel i and waveband j have the relationship as

$$DN_{ij} = T_i R_{ij} I_j \tag{3.11}$$

Terrain factor T_i only has relationship with azimuth and slope at a certain pixel i. T_i should be the same at all of wavebands. Illumination factor I_j describes the effect by solar radiation spectrum and it is the same at all of the pixels in a certain waveband. We assume that L_i is the geometric mean of DN_{ij}, $L_{\cdot j}$ is the geometric mean of all DN_{ij} in waveband j. $L_{\cdot\cdot}$ is the geometric mean of DN_{ij} of all the pixels. So the logarithmic residuals can be represented by [1].

$$Y_{ij} = (DN_{ij}/L_{i\cdot}) \cdot (L_{\cdot j}/L_{\cdot\cdot}) = (R_{ij}/R_{i\cdot}) \cdot (R_{\cdot j}/R_{\cdot\cdot}) \tag{3.12}$$

Logarithmic residuals eliminate terrain illumination factor and only corresponding with reflectivity.

(C) Empirical Linear Method

The earth surface is assumed to be Lambertian and atmospheric effects has already been considered. The DN values received by sensor consist of the followings [1]:

$$DN = L_s + L_0 + L_D \tag{3.13}$$

$L_s = L_g \ t$ = Ground radiation × (atmospheric transmittance + instrument gain parameters); L_0 = Path radiation + Dark current; L_D is the cross radiation which is due to atmospheric effects. The result is making image blur and the spectral behavior of near pixel is mixed into of pixel spectral response of the target. Effect of cross radiation is localized and can usually be eliminated by acquiring the atmospheric point spread function. For the point spread function $h(x, y)$ and an ideal image $f(x, y)$ without cross radiation, the image $g(x, y)$ under the effect of radiation is given by

$$g(x, y) = f(x, y) \oplus h(x, y) \tag{3.14}$$

The effect caused by path radiation and instrument drift should be corrected and the cross radiation of image DN' value should also be eliminated.

$$DN' = L_s + L_0 = L_g t + L_0 \tag{3.15}$$

Ground radiance is assumed to have a linear relationship between reflectivity.

$$L_g = kR + b_0 \tag{3.16}$$

According to the two equations above:

$$DN' = kR + b \tag{3.17}$$

The coefficients k and b can be calculated by the least squares if spectral response DN' of two or more kinds of feature objects and reflectivity R are known. The relationship between DN' and R can also be acquired [2]. We need spectral data in the field instead of extra atmospheric parameters and accurate atmospheric model. The absolute reflectivity is acquired, which is easy to operate and combine with standard database to recognize the objects.

3.5.2 Polarization Inversion Based on Empirical Linear Method

In (3.9), the Stokes parameters of an incident light have a linear relationship with imaging intensity. So a mathematical model of polarization inversion can be acquired according to the principle of empirical linear method. The data acquired by the polarization spectrometer are assumed to reflect the real polarization of the target and provide the priori multi-band polarization data of typical areas, so this method is suitable for multi-band polarization curve inversion. DN_0, DN_{45}, and DN_{90} are used to represent the DN values at $0°$, $45°$, and $90°$ of every waveband. The transmittance varies in different polarization directions according to the effect of polarization devices.

$$DN_0 = L_{s_0} + L_{o_0} + L_{D_0}$$
$$DN_{45} = L_{s_{45}} + L_{o_{45}} + L_{D_{45}} \tag{3.18}$$
$$DN_{90} = L_{s_{90}} + L_{o_{90}} + L_{D_{90}}$$

Collected images are clear in actual detection that has not involved aerial imaging scenes. The images have high spatial resolution. The effect between adjacent pixels is little so that the cross radiation $L_{D_0}, L_{D_{45}}, L_{D_{90}}$ can be ignored. The equation $L_s = L_g t$ is used in (3.18) that

$$DN_0 + DN_{90} = (L_{g_0} + L_{g_{90}})t_1 + (L_{o_{90}} + L_{o_0})$$
$$DN_0 - DN_{90} = (L_{g_0} - L_{g_{90}})t_2 + (L_{o_{90}} - L_{o_0}) \tag{3.19}$$
$$2DN_{45} - DN_0 - DN_{90} = (2L_{g_{45}} - L_{g_0} - L_{g_{90}})t_3 + (2L_{o_{45}} - L_{o_{90}} - L_{o_0})$$

$L_{g_0}, L_{g_{45}}$ and $L_{g_{90}}$ represent ground radiation intensity in the direction of $0°$, $45°$, and $90°$; t_1, t_2 and t_3 represent the effect of atmospheric transmittance and instrument gain in the three polarization directions. The polarization parameters from polarization spectrometer are assumed to be reliable and accurate. In (3.19), $(L_{g_0} + L_{g_{90}})$, $(L_{g_0} - L_{g_{90}})$, and $(2L_{g_{45}} - L_{g_0} - L_{g_{90}})$ represent actual polarization parameters in spectrometer sampling area are substituted by parameters data S_0, S_1, and S_2. The unknown t_0, t_{45}, t_{90}, $(L_{o_{90}} + L_{o_0})$, $(L_{o_{90}} - L_{o_0})$, and $(2L_{o_{45}} - L_{o_0} - L_{o_{90}})$ are calculated by the least squares method. These constant values obtained by the binary regression are applied to other pixels to obtain accurate multi-band polarization curves.

3.6 Experiment Settings

The imaging scene in the experiments was chosen as under direct sunlight at noon for accurate results. Special boards are used to simulate actual objects, polarization spectrometer, and multi-band polarization imaging system are used to collect multi-band polarization data in a vertical incident way to keep the lenses in a same direction. Figure 3.14 shows an experimental scene. Five different boards are on flat grass as samples, a black plank, a black tile, a green aluminum plate, an ordinary plank, and a grey iron plate, clockwise starting from the left corner. Radiation intensity values of each board were measured by the spectrometer.

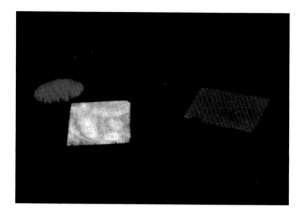

Fig. 3.14 Experiment setting for multi-band polarization imaging

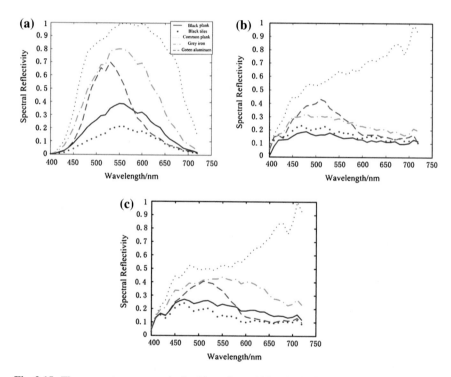

Fig. 3.15 The parameter curve s_0 obtained by **a** the multi-band polarization imaging system, **b** the spectrometer and **c** inversion

(1) Taking a small area on each sample board, the polarization parameters obtained from the image have a linear relationship with the polarization parameters obtained by spectrometer. Figures 3.15, 3.16 and 3.17 show

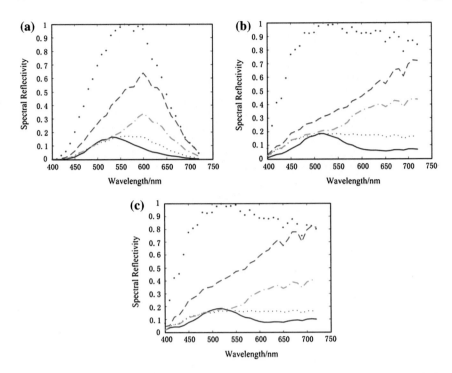

Fig. 3.16 The parameter curve s_1 obtained by **a** the multi-band polarization imaging system, **b** the spectrometer and **c** inversion

normalized multi-band polarization curves at every pixel in the experiment scene acquired by the least squares method. Figure 3.17 shows that the curve after inversion is consistent with the one obtained by spectrometer in the inversion. The accuracy of data model collected by the spectrometer and the multi-band polarization imaging system has already been verified. The accuracy of inversion curve is mainly influenced by the calibration accuracy and consistency of lens direction between the spectrometer and the multi-band polarization imaging system. Incorrect calibration will influence the correlation among multi-band polarization images, which makes the DN curve of multi-band polarization images in every direction deform so that the linear data model between the multi-band polarization imaging system and spectrometer may be corrupted. As for the anisotropy objects, the multi-band polarization state of their reflect light are different in every zenith and azimuth. The DN value and DN curve also greatly change so that the relationship at the condition of different zenith and azimuth are described as polarization bidirectional reflectance distribution function (P-BRDF). This model is not linear, however, so the shape of inversion curve and multi-band polarization data curve will be different due to the lens inconsistency of detection system and the spectrometer.

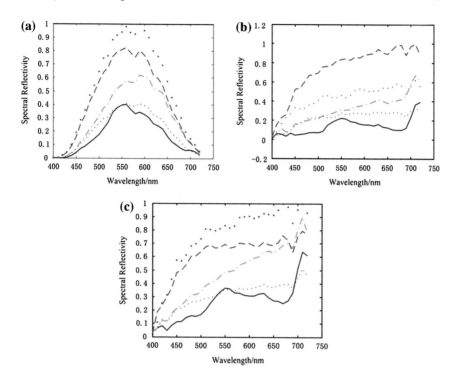

Fig. 3.17 The parameter curve s_2 obtained by **a** the multi-band polarization imaging system, **b** the spectrometer and **c** inversion

(2)   We use the spectral data as samples from four of these five boards to obtain relevant parameters following wave data, which is used to calibrate the spectrometer data in the other plate. Figure 3.18 shows the comparison of parameter curves acquired by spectrometer (dotted) and parameter curves acquired by multi-band polarization imaging system (solid). The data of black plank are considered as test samples and not used in the least-squares calculations. Calibrating the test samples using the fitting result of other four plates can achieve better results. The fitting results can be used to calibrate multi-band polarization regression line. Table 3.4 shows the values of k and b estimated using parameters inversion at different wavebands after classical linear inversion. Tables 3.4, 3.5 and 3.6 show that the changes of k and b are smooth near 550 nm. The closer to IR and UV wavelengths, the larger the changes. The detector is shown to be most stable near the green band. The transmittance of the detection system is relatively small and has lower SNR in the NIR and UV ranges, which will have greater impact on inversion.

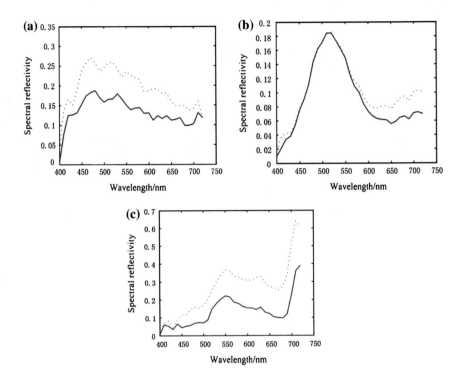

Fig. 3.18 The comparison of parameter curves **a** s_0, **b** s_1 and **c** s_2 obtained by spectrometer and calibration

Table 3.4 k, b values acquired by s_0 inversion

Wavelength (nm)	420	450	480	510	540	570	600	630	660	690	720
k	13.66	3.88	1.8	1.29	1.05	0.93	0.94	1.29	2.21	4.51	16.5
b	2908	556	430	−185	−120	290	988	2119	3281	4125	7276

Table 3.5 k, b values acquired by s_1 inversion

Wavelength (nm)	450	480	510	540	570	600	630	660	690
k	0.017	0.008	0.008	0.009	0.009	0.01	0.013	0.018	0.03
b	726	720	608	218	247	33	103	120	235

Table 3.6 k, b values acquired by s_2 inversion

Wavelength (nm)	420	450	480	510	540	570	600	630	660	690	720
k	0.45	0.14	0.07	0.06	0.04	0.04	0.05	0.05	0.1	0.18	0.56
b	219	28	−21	−126	−22	−23	−72	124	−18	240	754

References

1. Wang P, Zhenping B, Xiangnan L (2001) Remote sensing introduction. Higher Education Press, Beijing
2. Zhao Y (2004) Multi-source image fusion research based on polarization and spectral characteristics, School of Automation, Northwestern Polytechnical University, Xi'an, China
3. Wang D (2007) Imaging spectropolarimetric system and its applications in land cover classification, School of Automation, Northwestern Polytechnical University, Xi'an, China
4. Alouini M, Goudail F, Refregier P, Grisard A, Lallier E, Dolfi D (2004) Multispectral polarimeteric imaging with coherent illumination: towards higher image contrast. Proc SPIE 5432:134–144
5. Gupta N (2002) Remote sensing using hyperspectral and polarization images. Proc SPIE 4574:184–192
6. Gupta N, Dahmani R, Bennett K, Simizu S, Suhre DR, Singh NB (2000) Progress in AOTF hyperspectral images. Proc SPIE 4054:30–38
7. Shingu H, Homma K, Kurosaki H, Suzuki T, Yamamoto H (2002) Development of an onboard spectro-polarimeter for earth observation at NAL. Proc SPIE 4669:89–97
8. Homma K, Shingu H, Yamamoto H, Kurosaki H, Shibayama M (2004) Application of an imaging spectropolarimeter to agro-environmental sciences. Proc SPIE 5234:639–647
9. Yang Z (2003) Measurement and development of imaging spectropolarimetric. Chin J Spectrosc Lab 6(20):815–820
10. Zhao K, Zhong X (2001) Optics. The University Publishing Press, Beijing

Chapter 4
Multi-band Polarization Bidirectional Reflectance Distribution Function

Conventional remote sensing devices always identify a target by its reflection intensity or radiation intensity. While the polarization characteristics are always treated as noise. In the recent years, with a development of spatial remote sensing technologies, researchers have noticed the polarization characteristics which are determined by surface optical features in the process of reflection, scattering, and transmission. Measuring the polarization intensity, polarization degree, polarization angle, polarization ellipticity and emissivity can provide higher accuracy than radiation measurements. Generally, the polarization reflection spectrum contains different polarization states of different objects or different object states (such as medium, structure, surface roughness, and soil water content) which are closely related to wavelength. Polarization imaging can eliminate background noise and enhance the accuracy of target detection and identification, so that it has extensive application prospects in the field of target detection, classification, recognition, ocean remote sensing, image enhancement in severe weather conditions, computer vision, surveillance, intelligent transportation, and medical diagnostics. In this chapter, the polarization Bidirectional Reflectance Distribution Function (BRDF) is taken as an example to show the polarization characteristics of both target and background.

4.1 Bidirectional Reflectance Distribution Function

A bidirectional reflectance distribution function is an elementary optical characteristic which describes the reflection characteristics of target surface. Proposed in the 1970s by Nicodemus [1], BRDF describes the distribution of reflectance spectrum to provide spatial information in the reflectance spectrum [2]. Defined and developed in terms of radiation, BRDF theory has been widely applied to scattering and radiation problems in the visible, infrared, and the microwave spectra. BRDF is a basic optical parameter describing the radiation energy distribution of light in a

© National Defense Industry Press, Beijing and Springer-Verlag Berlin Heidelberg 2016   73
Y. Zhao et al., *Multi-band Polarization Imaging and Applications*,
Advances in Computer Vision and Pattern Recognition,
DOI 10.1007/978-3-662-49373-1_4

certain incident direction after surface reflection [1]. BRDF has been applied in target segmentation using the scattering properties in the visible light, infrared, and microwave radiation. When the light is irradiated to a target surface, the measured light is divided into surface scattering light and volume scattering light components. Scattering characteristics are closely related to roughness of a surface. For a smooth surface, specular reflection component becomes strong and the diffusive scattering component is weak. If the target is a mixture of different materials, the volume scattering is caused by multi-scattering by a part of the transmittance light. A BRDF of the target surface is developed for quantifying the difference of surface scattering and volume scattering in different incident and detection directions. BRDF describes spatial scattering characteristics. A study on the BRDF of coated metal reveals the volume scattering effect, while the transmittance and volume scattering can be ignored since uncoated metal has a large dielectric constant. For a target such as soil and vegetation, surface scattering and volume scattering should not be ignored, and the volume scattering plays an important role in spatial scattering characteristics [3, 4]. A general polarization BRDF does not only quantify the scattering direction, but also describes scattering polarization characteristics in the applications of atmospheric detection, land cover classification, environmental monitoring, crop analysis, medical diagnostics, and military target detection. Figure 4.1 shows that a target object can be determined by the surface roughness, dielectric constant, radiation wavelength and polarization, and the geometric relationships. In Fig. 4.1, dA denotes an incremental piece of the surface, (θ_i, ϕ_i) is the incident direction, (θ_r, ϕ_r) is the observation direction of detector, where the subscripts i and r represent incident direction and reflection direction respectively,

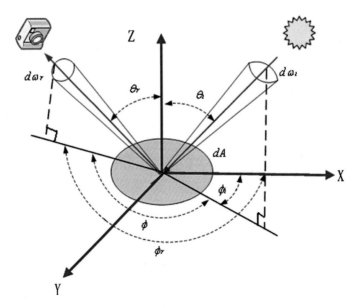

Fig. 4.1 Geometric relationship definition for BRDF

θ is the zenith angle and ϕ is the azimuth angle. Z denotes the average normal direction of a rough surface.

A BRDF is defined as the ratio of the emitted radiation $dL_r(\theta_i, \phi_i, \theta_r, \phi_r)$ along the emission direction and the irradiance $dE_i(\theta_i, \phi_i)$ of surface along the incident direction:

$$f_r(\theta_i, \theta_r, \phi, \lambda) = \frac{dL_r(\theta_i, \theta_r, \phi)}{dE_i(\theta_i, \phi_i)} \qquad (4.1)$$

where λ denotes the wavelength and $\phi = \phi_r - \phi_i$. Since the radiance is defined as the radiant flux $(W/(m^2 \cdot sr))$ (W means watt and sr means steradian) of unit area and unit solid angle along the radiation direction. The irradiance denotes the radiant flux of unit area so that the unit of BRDF is sr^{-1}.

4.1.1 Measurement of Bidirectional Reflectance

Research on BRDF has been conducted since 1980s in the countries including the United States, United Kingdom, and France. To acquire a BRDF database of targets and their backgrounds, researchers developed various measurement methods [5–7]. With a wide use of multi-band polarization information in remote sensing, medicine, target detection and recognition, researches on multi-band polarization BRDF have drawn significant attention. The instruments to measure BRDF include field spectrometers such as Analytical Spectral Device (ASD) and GER spectrometer, which are used to acquire a wide spectral range and accurate radiation information. Spatial imaging CCD is also applied with LCTF or AOTF to acquire polarization BRDF in all wavebands, the main advantage of this method is acquiring the spatial information of target, but the accurate radiation is not available because of the noises, exposure time and dark current. Moreover, spectrum range is also limited by the filters. As most hyperspectral sensors have a spectral resolution of 10 nm [8, 9], the spectral resolution in BRDF measurement is always taken as 10 nm. Multi-angle observation is a key technology of multi-band polarization BRDF measurement. Common devices measure the angles of mechanical rotation and optical mapping to realize multi-angle observation [9].

Mechanical rotation devices can achieve spatial hemisphere BRDF sampling by rotating around a vertical axis and a horizontal axis. In 1995, the Remote Sensing Laboratory of University of Zurich constructed the Dual-view Field-goniometer System (FIGOS) to measure BRDF. The system contains the zenith arched, orientation and mobility slide rails. It can collect hyperspectral BRDF data in the spatial range of zenith $[-75°, 75°]$ and azimuth $[0°, 360°]$, which is shown in Fig. 4.2a [10]. Meanwhile, the experiment has also built the Laboratory-Goniometer System (LAGOS) in Fig. 4.2b [11]. LAGOS uses the GER100 spectrometer as detector and can only acquire the radiation of point, but cannot acquire spatial information about experimental scenes or targets. All of these devices can accurately measure the

(a)   (b)

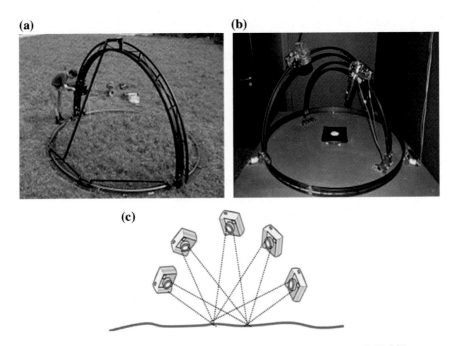

(c)

Fig. 4.2 Three BRDF measurement method. **a** FIGOS. **b** LAGOS. **c** Narrow FoV CCD

hemisphere indoor or outdoor with high angular resolution and fast acquisition speeds. James R. Shell II from the Digital Imaging and Remote Sensing (DIRS) Laboratory, Rochester Institute of Technology proposes an outdoor multi-band polarization BRDF measurement using a tripod and a polarization CCD in 2005, a moving CCD is used to realize multi-angle collection [12]. The system is simple, economical, practical, and accurate, but only acquires multi-band polarization BRDF data at some certain points in hemisphere space, as shown in Fig. 4.2c, where the CCD has a small FOV and the acquired images are usually considered to have the same reflection angle [9].

Optical mapping apparatus reflects light in the whole hemisphere space into CCD sensor by oval concave mirror and plane mirror. The schematic diagram of BRDF measurement is respectively shown in Fig. 4.3a. As shown in Fig. 4.3b, this CCD has large FOV and collects points with different reflection angles at the same time, then the scattering angle will be converted into sensor coordinate system by angle mapping function to achieve BRDF data collection. Optical mapping apparatus can collect accurate and high-angle resolution BRDF data without any moving parts. Comparing with mechanical rotation, it has higher collection speed and accuracy. This method saves acquisition time of multi-angle BRDF, but has a disadvantage that the geometrical structure of a target surface should be known accurately. Every pixel in acquired images corresponds to a BRDF measurement value in different angle which is one to one with geometry of the target.

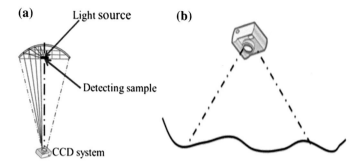

Fig. 4.3 Optical mapping apparatus

A BRDF can be measured both indoor and outdoor. Indoor measurement technology is relatively mature, usually the measuring instruments include a light source, detector and angle measuring devices. Most experiments take fixed light source positions and change azimuth and zenith angles of the upper hemisphere space on the target sample by measuring the angle of the detector device. The light source mainly uses halogen, laser source or some other wider wavelength range and uniform light source so that we can use spectral filter to choose wavelength and acquire multispectral or hyperspectral BRDF data. Comparing with the measurements in laboratory, the outdoor condition is more complex because of the difficulties caused by spatial inhomogeneity of natural material and weather condition. Outdoor measurement mainly concludes ground measurement and aerospace remote sensing measurement. And the ground measurement can be divided into fixed sensor and moved sensor measurement [13–24]. Figure 4.4 shows two commonly used multi-band polarization BRDF imaging systems that contain a measurement system on ground and aviation platforms.

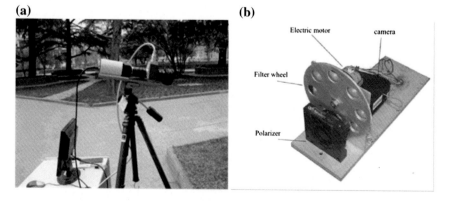

Fig. 4.4 a Multi-band polarization BRDF imaging system on ground. **b** Multi-band polarization BRDF imaging system on aviation platforms

Fig. 4.5 Multi-angle measurement system of China. **a** Rotating bracket multi-angular viewing equipment. **b** Bidirectional reflectance measuring device, **c** outdoor multi-angle BRDF measurement system

The Remote Sensing and GIS Research Center of Beijing Normal University has cooperated with Institute of Physics of Remote Sensing for a study of BRDF since late 1990s. They built rotating bracket multi-angular viewing equipment shown in Fig. 4.5a and Airborne Multi-angle TIR/VNIR Imaging System (AMTIS). The outfield experiment system works on main plane and cross main plane in the zenith range of [0°, 65°] (with an interval of 5°) and the aviation multispectral imaging system can collect multi-band BRDF data of 9 detection angles in the visible, near infrared, and thermal infrared bands. The feature bidirectional reflectance measuring device developed by Changchun Institute of Optics machinery is shown in Fig. 4.5b. It consisted of three parts including light source system, bidirectional reflectance photometer system and control system. The photometer has two wavebands A (630–690 nm) and B (760–1100 nm), azimuth 0°–350°. The nickel-tungsten light source has an incident direction in the range of 0°–60° with a step of 10°. There are 7 probes on the detection frame where the interval between each probe is 10° and the detection range is 0°–60°. Meanwhile, it comes with polarized lenses that can rotate at any angle and measure the reflectance radiation all around from multiple observation points. Fast data acquisition and processing are completed by electronic control panel and micro-computer, the data are output in the form of a table or a graph to make statistical analysis. Figure 4.5c shows an indoor BRDF laser measurement system and the outdoor multi-angle BRDF measurement system developed by the Anhui Institute of Optics and Precision Mechanics and Xidian University in 1990s. It has the similar multi-angle implementation with FIGOS, including an automatic measuring frame and two wild type spectrometers. One spectrometer is fixed on the frame while the other is placed on the ground to measure the radiation of diffusive reflector. The measurement range of zenith and azimuth are [−75°, 75°] and [0°, 360°], respectively.

4.1.2 Bidirectional Reflectance Model

Apart from the experiment method, BRDF also can be acquired by BRDF modeling. Research on BRDF modeling has already achieved many results in the field of quantitative remote sensing, machine vision and computer graphics. Hundreds of theoretical models are established at present in three different genres:

- Statistical empirical model
- Physical model
- Empirical/semi-empirical model

The physical model of BRDF is the main interest of BRDF study. It has achieved the most comprehensive theoretical basis. The model parameters have clear physical meanings and mathematical description of the mechanism. Other possible models are:

- Radiation transfer model
- Geometrical optics model
- Hybrid model
- Computer simulation model

Radiation transfer function with multi-scattering effect is the theoretical foundation of a radiation transfer model, especially in the infrared and microwave regions. But the three-dimensional differential equations are so complexed that can only be solved numerically without considering BRDF anisotropy. Applicability of geometrical optics model is broader, which mainly considers macroscopic geometry of the features. The features are assumed to have known geometry, optical properties, and arrangements. Following the geometrical optics theory, we analyze intercepted and cover of incident light and the reflection of the earth surface to ensure the directional reflection in vegetation canopy. But a basic ray model is not reliable when the size of small patch is smaller than or proportional to the wavelength. The computer simulation model is based on the statistical Monte-Carlo method where the given distribution function are applied. And to some extent, the computer simulation model can also be used as a validation tool for other models. However, the disadvantages are obvious that it costs lots of sources and time and the probability density function should be predicted. The semi-empirical model focuses on the main effect factors of BRDF, utilizing the measurement data as well as basic physical principles to make it intuitive, simple, effective, and promising. At present, the semi-empirical model has developed to simulate different targets and scenes [2, 8].

An initial Lambert model is designed to describe an ideal diffuse surface where the scattering properties are independent with the observation position. However, an ideal Lambert model does not exist. A semi-empirical model of the background material such as grass, soil and the desert commonly uses a kernel model. The model contains surface scattering kernels, volume scattering kernels and diffuse scattering kernels to simulate vegetation BRDF models. A soil model is mainly based on the Hapke model [25], Phong model [26], Walthall and their developed models [27]; while the Otterman model [28] and Staylor-Suttles model [29] are applied to simulating soil, desert and arid regions. Background models are effectively used in the process of remote sensing data. And the Ward model is pretty efficiently to simulate the surface scattering properties of metals and other anisotropy materials. The Torrance-Sparrow model [30] and Cook-Torrance model [31] are used in a wide range of computer graphics, they have V-shaped groove structure, shadowing and concealing effects so that they can simulate the targets with V-shaped structure. Torrance-Sparrow model is fitted with non-specular peak scattering properties while the Cook-Torrance model can simulate the scattering properties of both metals such as copper and gold, and non-metallic materials such as the plastics and ceramics. Apart from these, BRDF is also described as the different linear combinations of wavelets, spherical harmonics, Zernike and other basic functions which are related to wavelength and appear as different shapes. The Maxwell-Beard model [32] and the Sanford-Robertson model [33] are used to simulate the coated surface and aerial infrared signals at the very start. The five-parameter model is established to simulate the reflection of rough surface. Some newly developed coated target models are designed to divide the reflection into two function terms with different reflection coefficients, the specular and diffuse reflection. It is helpful for the inversion through different parameter inversion methods.

The models discussed above are all concerning a scalar BRDF model without the change of polarization caused by target and background. Both of the intensity and polarization should be taken into account when incident light acts on the surface of target. Priest [34] has established a polarization BRDF model with the start of microfacet model, shadowing functions and Fresnel reflection. In this model, Fresnel reflection is converted to the polarized matrix through Jones matrix and coordinates transformation, then various small models are combined to make a multi-band polarization BRDF model. The parameters of this model such as the surface roughness and refractive index are different in different bands. The model is complex and the parameter inversion is difficult if the effects of shadow and cover among microfacet elements are considered [35–38].

In outdoor remote sensing, the intensity of multi-band polarization BRDF (scalar BRDF), the DoLP and *Orient* are used to achieve multi-band polarization BRDF modeling. The intensity model is known as a scalar BRDF model. The approximate polarization components model is established by analyzing the relationship between polarization components and probe geometry such as incident sunlight angle, detection angle, and azimuth angle. The polarization angle component model is established by analyzing the reflection of small patches and coordinate transformation. This method completely relies on experimental data, which is simple and feasible.

4.2   Polarization BRDF

Light is a transverse wave that the light vector is perpendicular to the propagation direction. So the direction of light vector at any point and any time in light field can be specified to completely describe the light wave. Polarization is a performance of light vector property which reflects time-varying electric field vector magnitude and direction with the change of propagation direction. The polarization properties of reflection radiation are not only dependent with the observation geometric condition but also the nature of target and illumination effects. Outdoor situation is more complicated for the reason that the radiation source is integrated by the sunlight, scattered light and reflected radiation. Artificial surface is substantially unnatural smooth surface and has higher degree of polarization than natural surface, while the radiation from rough surface is multiple scattering light and have smaller degree of polarization. The dark surface also shows greater degree of polarization, it is because the proportion of single scattering is large on dark surface, while the multiple scattering is more dominant on bright surface [39]. Although the visibility of intensity image proves to be better than polarization image, but the latter has higher contrast and excavates many hidden information in intensity image to identify camouflage and efficiently enhance the edge of target [40].

4.2.1   Polarization BRDF Functions

Polarization BRDF (P-BRDF) is a more general form of scalar BRDF. It has a complete description about reflection characteristic of materials which not only quantifies the values of scattering but also gives the scattering polarization characteristics. Only P-BRDF can show the value of total reflection coefficient correctly if the incident light is partly polarized. The P-BRDF is converted to Muller matrix when incident light and reflected light are represented by Stokes vector.

$$dL_r(\theta_i, \theta_r, \phi) = F_r(\theta_i, \theta_r, \phi, \lambda) \times dE_i(\theta_i, \phi_i) \tag{4.2}$$

Muller matrix is generally used to describe properties of the propagation medium. It is a 4×4 matrix without any unit and always be normalized, which means the first element is valued as 1 and constant coefficients out of matrix are ignored. This concept describes polarization properties of the medium with the cost of losing absolute radiation. But when the Muller matrix is used to describe BRDF, the unit of matrix is sr^{-1} and constant coefficients cannot be ignored. In this condition, the first element f_{00} in matrix F_r also equals with scalar BRDF f_r [35]. Suppose that there is no circularly polarized light reflected by nature surface, then Stokes vector contains only first three elements and the Mueller matrix will be reduced to a 3×3 matrix. For this 3D matrix and vector, the above polarization BRDF can be defined as

$$\begin{bmatrix} s_0^r \\ s_1^r \\ s_2^r \end{bmatrix} = \begin{bmatrix} f_{00} & f_{01} & f_{02} \\ f_{10} & f_{11} & f_{12} \\ f_{20} & f_{21} & f_{22} \end{bmatrix} \begin{bmatrix} s_0^i \\ s_1^i \\ s_2^i \end{bmatrix} \qquad (4.3)$$

where s_0^r, s_1^r and s_2^r are Stokes parameters of the reflected radiation, s_0^i, s_1^i and s_2^i are Stokes parameters of the incident irradiance [36, 37].

4.2.2 Polarized Radiative Transfer Function

Radiative transfer function describes the energy transmission process among sunlight—the surface of target—detector and the scattering absorption of sunlight in atmosphere. Polarized transfer function is a more general one that considers not only the change of energy in transmission, but also polarization state. Research on polarized transfer function helps us to understand the energy transmission in interaction process between sunlight and target, meanwhile it also provides theoretical basis for the quantitative analysis of polarization information in remote sensing.

As shown in Fig. 4.6, the surface of target accepts energy from direct sunlight $\mathbf{L}_r$ and atmospheric scattering $\mathbf{L}_s$ in visible remote sensing region. The total radiation is approximately divided into three parts:

(1) Sunlight $\mathbf{L}_r$ after scattering: It contains scattering energy of the target $\mathbf{L}_{rs}$, $\mathbf{L}_{rss}$ at the surface and internal. And $\mathbf{L}_{rss}$ occupies only a small proportion so it is seldom considered in the visible light.
(2) Air scattering light $\mathbf{L}_s$ reflected by the target.
(3) Sunlight $\mathbf{L}_u$ enters the director directly along the direction of target-detector after atmospheric scattering. It also contains a small energy proportion.

Fig. 4.6 Schematic diagram of the light received by a detector

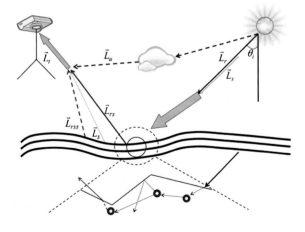

Regardless of atmospheric conditions and the reflection of the ground or target, the three kinds of radiation mentioned above are in descending order, and all of them are functions of incident angle, scattering angle and azimuth angle. First consider scalar quantity L_r and then spread it to vector form: assuming solar irradiance outside the atmosphere is E_s, which has transmittance ratio τ_i as transmitting along sunlight. Thus L_r can be represented:

$$L_r = \tau_r(\theta_r)f_r(\theta_i, \theta_r, \phi)\cos\theta_i\tau_i(\theta_i)E_s(\theta_i) \tag{4.4}$$

Muller matrix form $\mathbf{F}_r$ is used to instead of scalar quantity f_r. And E_s is in placed by Stokes vector $\mathbf{E}_s$.

$$\mathbf{L}_r = \tau_r(\theta_r)\mathbf{F}_r(\theta_i, \theta_r, \phi)\cos\theta_i\tau_i(\theta_i)\mathbf{E}_s(\theta_i) \tag{4.5}$$

Coordinate transformation should be applied in (4.4) and (4.5). As BRDF is identified in accordance with normal direction of target surface, which is also not consistent with zenith, we should transform the coordinate to keep them in the same direction. In this book, we only consider the situation that normal and zenith are in the same direction. In the same way, consider scalar quantity L_s and then spread it to vector form. Assuming $L_s^{\Omega_i}$ as the sky scattering light (Ω_i represents different positions), the total scattering light of target surface can be obtained by modifying cosine of incident angle, reflecting and integrating the sky scattering light in upper hemispherical space. Then, the total scattering light is weaken through transmission from target and detector and reaches at the detector as L_s.

$$L_s = \tau_r(\theta_r)\iint\limits_{\Omega_i} f_r(\theta_i, \theta_r, \phi)\cos\theta_i L_s^{\Omega_i}(\theta_i, \phi)d\Omega_i \tag{4.6}$$

Here $d\Omega_i = \sin\theta_i d\theta_i d\phi$. Muller matrix form $\mathbf{F}_r$ is used to instead of scalar quantity f_r. And $L_s^{\Omega_i}$ is in placed by Stokes vector $\mathbf{L}_s^{\Omega_i}$. So the corresponding vector representation of $\mathbf{L}_s$ is

$$\mathbf{L}_s = \tau_r(\theta_r)\iint\limits_{\Omega_i} \mathbf{F}_r(\theta_i, \theta_r, \phi)\cos\theta_i \mathbf{L}_s^{\Omega_i}(\theta_i, \phi)d\Omega_i \tag{4.7}$$

Expressions of L_u and its vector form $\mathbf{L}_u$ are quire complex so always be ignored in actual measurement. L_u and $\mathbf{L}_u$ are simplified by their geometrical relationship:

$$L_u = L_u(\theta_r, \phi) \tag{4.8}$$

$$\mathbf{L}_u = \mathbf{L}_u(\theta_r, \phi) \tag{4.9}$$

It has known that L_r and L_s are directly related with their scalar quantity f_r, and their vector forms are also related with corresponding Muller matrixes $\mathbf{F}_r$. As for ideal diffuse reflection surface, the scalar quantity f_r has no direction, i.e. it is independent with geometrical condition and the geometrical position of light source so can be instead by Lambertian surface BRDF ($f_r = \rho/\pi$). Atmospheric attenuation factor is only to be known when calculate L_r and L_s, which can extensively simplified the radiation equation. However, due to non-Lambertian characteristics showed by more than 99 % natural targets, they cannot be instead by f_r. Radiation Transfer function can be represented:

$$L_t = L_r + L_s + L_u$$
$$= \tau_r f_r \cos\theta_i \tau_i E_s + \tau_r \iint\limits_{\Omega_i} f_r \cos\theta_i L_s^{\Omega_i} d\Omega_i + L_u \tag{4.10}$$

The expression of f_r can be derived from (4.10):

$$f_r = \frac{L_t - \tau_r \iint_{\Omega_i} f_r \cos\theta_i L_s^{\Omega_i} d\Omega_i - L_u}{\tau_r \tau_i \cos\theta_i E_s} \tag{4.11}$$

The corresponding vectorial radiation transfer function is

$$\mathbf{L}_t = \mathbf{L}_r + \mathbf{L}_s + \mathbf{L}_u$$
$$= \tau_r \mathbf{F}_r \tau_i \cos\theta_i \mathbf{E}_s + \tau_r \iint\limits_{\Omega_i} \mathbf{F}_r \cos\theta_i \mathbf{L}_s^{\Omega_i} d\Omega_i + \mathbf{L}_u \tag{4.12}$$

Since the solar radiation in outer atmosphere is random polarized, we only need to consider its value of scalar quantity, which means the first element of Stokes vector. Thus, $\mathbf{E}_s$ is in place of E_s. And only the first row of polarization BRDF matrix is needed in the expression of $\mathbf{L}_r$. In passive remote sensing, the first row of polarization BRDF matrix can be only inversed, which is claiming as the Stokes vector representation of polarization BRDF. From (4.12),

$$\tau_r \begin{bmatrix} f_{00} \\ f_{10} \\ f_{20} \end{bmatrix} \tau_i \cos\theta_i E_s = \mathbf{L}_t - \tau_r \iint\limits_{\Omega_i} \mathbf{F}_r \cos\theta_i \mathbf{L}_s^{\Omega_i} d\Omega_i - \mathbf{L}_u \tag{4.13}$$

The Stokes vectorial representation of polarization BRDF is given by

$$\begin{bmatrix} f_{00} \\ f_{10} \\ f_{20} \end{bmatrix}_{(\lambda)} = \frac{\mathbf{L}_t - \tau_r \iint_{\Omega_i} \mathbf{F}_r \cos\theta_i \mathbf{L}_s^{\Omega_i} d\Omega_i - \mathbf{L}_u}{\tau_r \tau_i \cos\theta_i E_s} \tag{4.14}$$

It is the radiation transfer function of polarization BRDF that derived from (4.11).

4.3 Polarization BRDF Measurements

Polarization BRDF radiation transfer function is applied to obtain f_{00}, f_{10}, and f_{20} in BRDF matrix so that the polarization BRDF data of different materials can be acquired. Polarization BRDF measurement instruments are mainly polarization spectrometer and imaging polarization spectrometer. Polarization spectrometer has highly spectral resolution, measuring speed and accuracy, but the data can only described the average reflect characteristics so it is difficult to apply in classification and identification researches. The data measured by imaging polarization spectrometer can describe texture or variation of polarization BRDF, but it is hard to measure polarization BRDF data and describe polarization BRDF characteristics due to the serious influence of equipment response function. In order to make up these shortages, polarization BRDF data should be measured by the above instrument firstly, then the inversion of data should be done using polarization spectrometer, finally polarization BRDF characteristics and texture can be obtain precisely.

4.3.1 Measuring Principles

From (4.14), the radiation transfer function of polarization BRDF is acquired

$$
\begin{bmatrix} f_{00} \\ f_{10} \\ f_{20} \end{bmatrix}_{(\lambda)} = \frac{\mathbf{L}_t - \mathbf{L}_s - \mathbf{L}_u}{\tau_r \tau_i \cos \theta_i E_s} = \frac{\mathbf{L}_t - \tau_r \iint_{\Omega_i} F_r \cos \theta_i \mathbf{L}_s^{\Omega_i} d\Omega_i - \mathbf{L}_u}{\tau_r \tau_i \cos \theta_i E_s} \tag{4.15}
$$

$\mathbf{L}_t$ is the Stokes vector of total radiation. It can be measured under sunlight. $\mathbf{L}_s$ is the Stokes vector of sky scattering light reflected by target, it is measured in shadow. $\mathbf{L}_u$ is approximate to zero so that it can be negligible in ground remote sensing. Denominator parameters in (4.15) are unknown and difficult to measure, so it is necessary to get polarization BRDF by the comparative measurement method [38]. Assuming polarization BRDF of a certain reference material is

$$
\begin{bmatrix} f_{00\_ref} \\ f_{10\_ref} \\ f_{20\_ref} \end{bmatrix}_{(\lambda)} = \frac{\mathbf{L}_{t\_ref} - \mathbf{L}_{s\_ref}}{\tau_r \tau_i \cos \theta_i E_s} \tag{4.16}
$$

where $\mathbf{L}_{t\_ref}$ and $\mathbf{L}_{s\_ref}$ represent radiation intensity of the reference board in sunlight and shadow respectively. $E = \tau_r \tau_i \cos \theta_i E_s$, L_{t_ref} and L_{s_ref} are the first elements. From (4.16) we have

$$E = \frac{L_{t\_ref} - L_{s\_ref}}{f_{00\_ref}} \tag{4.17}$$

So here is the polarization BRDF of target in the same condition.

$$\begin{bmatrix} f_{00} \\ f_{10} \\ f_{20} \end{bmatrix}_{(\lambda)} = \frac{L_t - L_s}{E} = \frac{f_{00\_ref} \times (L_t - L_s)}{L_{t\_ref} - L_{s\_ref}} \tag{4.18}$$

Obeying the comparative measurement method, f_{00_ref} of reference material needs to be known in the upper hemispherical. While, there is only the ideal diffuse reflection board has fixed BRDF equaled ρ/π (ρ represents the hemispherical reflectivity which is measured through integrating sphere) [38]. In actual experiments, standard white board is used as the reference due to its better diffuse reflection, high reflectance ratio, small spectral selectivity and high stability. As the low proportion of elliptical polarization in remote sensing, so the intensity of Stokes vector components can also be approximated into the addition of two orthogonal linear polarization intensities.

$$s_0 = I_0 + I_{90} \tag{4.19}$$

So the radiation intensity of reference board in sunlight and in shadow can be respectively shown as follows:

$$L_{t\_ref} = I_{t\_ref0} + I_{t\_ref90} \tag{4.20}$$

$$L_{s\_ref} = I_{s\_ref90} + I_{s\_ref90} \tag{4.21}$$

In theory, the calibration images can be doubled to obtain the target BRDF if radiation is non-polarized. However, the reference board may exhibit less polarization characteristics, so images in two orthogonal polarized directions are needed to calibrate the measurement system precisely. Comparing calculation formulas of L_t, L_s and Stokes vector, inserting (4.20) and (4.21) into (4.18), we obtain the Stokes vector calculation formula of target BRDF.

$$\begin{bmatrix} f_{00} \\ f_{10} \\ f_{20} \end{bmatrix}_{(\lambda)} = \frac{\rho}{\pi(I_{t\_ref0} + I_{t\_ref90}) - (I_{s\_ref0} + I_{s\_ref90})}$$

$$\times \begin{bmatrix} ((I_{t\_0} - I_{s\_0}) + (I_{t\_45} - I_{s\_45}) + (I_{t\_90} - I_{s\_90}) + (I_{t\_135} - I_{s\_135}))/2 \\ (I_{t\_0} - I_{s\_0}) - (I_{t\_90} - I_{s\_90}) \\ (I_{t\_45} - I_{s\_45}) - (I_{t\_135} - I_{s\_135}) \end{bmatrix} \tag{4.22}$$

As for the polarization BRDF data of single spectral band in a certain detecting position, at least four groups of reference data and eight groups of target data are needed. Similarly, DoLP and Orient are also derived:

$$DoLP = \frac{\sqrt{f_{10}^2 + f_{20}^2}}{f_{00}} \tag{4.23}$$

$$Orient = \frac{1}{2}\tan^{-1}\left(\frac{f_{20}}{f_{10}}\right) \tag{4.24}$$

4.3.2 Outdoor Polarization BRDF Measurement Systems

Multi-band polarization imaging system shown in Fig. 3.9, mainly consists of objective lens, LCTF, controller, Relay Lens, and Qimage 14 bit high precision camera. All the components are integrated on the dock and charged by portable computer and battery. The structure is compact and dexterous, especially suitable for field operations. The advantages in response time and weight make it helpful in aerial operation on UAV. DN value is used to represent target radiation. Considering the effect caused by dark current and imaging environment, GER 2600 spectrometer is applied in the system to obtain the surface radiation, spatial and multi-band polarization information. The VariSpec LCTF is used as filter. It is faster and more precise than mechanical runner filters. Figure 3.10 shows the spectral characteristics. LCTF is placed in front of imaging lens, operating in visible light and near-infrared bands. LCTF controller is connected to the computer through USB, where the spectral resolution can be adjusted in software.

LCTF is a new spectroscopic device made according to electrically controlled birefringence effect of liquid crystal. It has light modulation, polarization, filter and other functions which are beyond the optical devices in traditional spectral analysis technique such as grating and prism. It has totally 6 Lyot filters in cascade. The principle of every Lyot unit is shown in Fig. 3.3. Every Lyot slide unit is consisted of polarizer, phase retarder and liquid crystal. Spectral response curve of VariSpec LCTF in 400–720 nm is shown in Fig. 3.11. Transmission is less than 30 % when wavelength is less than 500 nm, so imaging quality is low and degradation effects become more obvious.

4.3.3 Imaging Data Inversion

Received signals interfere with atmospheric scattering, target surface, terrain and radiation spectrum. Atmospheric interference should be eliminated through polarization inversion primarily. Empirical linear inversion only acquires data in the wild

without additional atmospheric parameters and accurate atmospheric model. While
a spectrometer can exactly provide multi-band polarization data, so empirical linear
inversion is a good option to get image data inversed. Assuming the ground is
approximately Lambert and taking the atmospheric effect into account, the DN
value received by sensor consists of the three parts:

$$DN = L_1 + L_2 + L_3 \qquad (4.25)$$

where $L_1 = L_g \times t =$ ground radiation $\times$ (atmospheric transmittance + detector's gain
parameter). L_2 represents atmospheric path radiation and dark current, L_3 represents
cross radiation. Cross radiation due to atmospheric scattering that light path connected target element and its neighbor elements is received by the sensors, the result
of which are image blurring and spectral mixing. However, it has little influence in
ground remote sensing and therefore can be ignored. So here we only need to
calibrate atmospheric path radiation and device drift. (4.25) can be rewritten as:

$$DN = L_1 + L_2 = L_g \times t + L_2 \qquad (4.26)$$

As the interference of CCD and LCTF response function, the data cannot precisely describe real reflection spectral characteristics. Using DN_0, DN_{45}, DN_{90} and
DN_{135} represent DN values in polarization directions of $0°$, $45°$, $90°$ and $135°$, L_g
represents the spectrometer data of corresponding polarization directions. From
(4.26) we can get inversion formula of polarization empirical linear inversion.

$$\begin{bmatrix} DN_0 \\ DN_{45} \\ DN_{90} \\ DN_{135} \end{bmatrix}_{(\lambda)} = \begin{bmatrix} L_{g0} \times t_0 \\ L_{g45} \times t_{45} \\ L_{g90} \times t_{90} \\ L_{g135} \times t_{135} \end{bmatrix} + \begin{bmatrix} L_0 \\ L_{45} \\ L_{90} \\ L_{135} \end{bmatrix} \qquad (4.27)$$

If the DN values and L_g of two or more objects are known, the coefficient t and
L can be solved through least squares method. And relationship between DN values
and L_g is acquired by feeding t and L in (4.27) [38].

4.3.4 Experiment Procedures

Figure 4.7 shows the procedure of polarization BRDF acquisition, and the detailed
explanations are presented as follows:

(1) Install and debug: Install the system and target according to experimental
    purpose and scenes (target, azimuth and zenith of detector). Meanwhile, adjust
    exposure time and gain according to sunlight.
(2) Measure radiation calibrating: Collect multi-band images of calibrating board
    and spectrometer data of $0°$ and $90°$ respectively in sunlight and shadow.

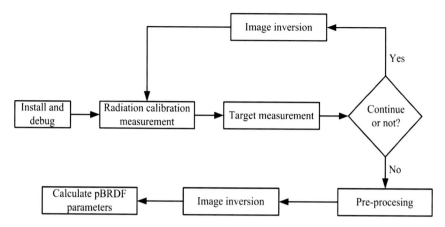

Fig. 4.7 Polarization BRDF acquisition procedure

(3)   Measure target: Collect the multi-images of target and spectrometer data of 0°, 45°, 90° and 135° respectively in sunlight and shadow.
(4)   Renew the location of collecting system or experimental scenes. Repeat steps (2) and (3).
(5)   Pre-processing. Due to translation deviation among different polarization degrees, images should be registered and enhanced or deblurred.
(6)   Image inversion. Choose two kinds of material in experimental images (To reduce the error, we choose the white object and black body), establish the relationship between multiple bands images and spectrometer data by least square method and inverse data of other materials.
(7)   Calculate polarization BRDF parameters. Calculate Stokes vector parameters of polarization BRDF, DoLP and Orient using (4.22)–(4.24).

More than 20 groups of images are collected in different scenes. These images are considered as a polarization BRDF image database which is consisted of polarization BRDF characteristics, image classification and model studies. In every data group, there are radiation calibration data, target data, spectrometer data and corresponding details.

4.3.5   Simulation Results and Analysis

The following experiments are designed for comparing polarization BRDF before and after inversion. We have collected polarization BRDF images and spectrometer data for white tiles, wooden board and black rubber. Then white tiles and black rubber are used as white object and black body to inverse image data of wooden board through empirical linear inversion. Figure 4.8 shows the polarization BRDF

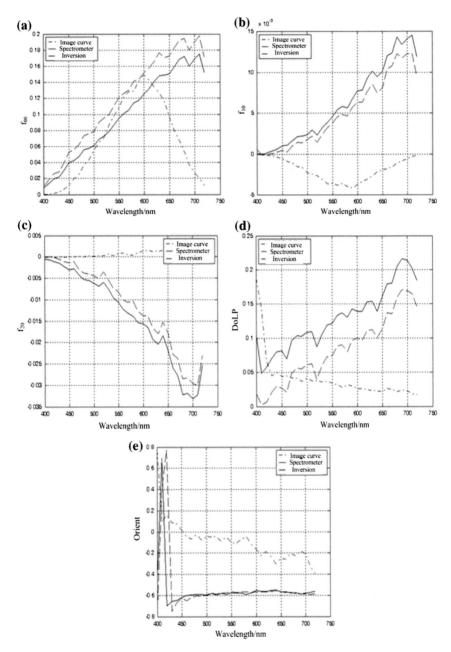

Fig. 4.8 comparison of polarization BRDF curve before and after inversion. **a** f_{00} curves. **b** f_{10} curves. **c** f_{20} curves. **d** *DoLP* curves. **e** *Orient* curves

parameter curve of wooden board (curve 1), spectrometer data (curve 2) and inversion data (curve 3).

CCD and LCTF have better transmittance in middle bands than edge bands, so curve of f_{00} shows the parabolic shape that the middle is higher than both sides. This kind of interference can be eliminated by the empirical line inversion. We can see that every polarization BRDF curve showing big differences with its corresponding spectrometer curve, and inversion data is close to spectrometer data. As for the curves of f_{00}, f_{10}, f_{20} and *Orient*, the average errors are less than 12 % and only the curve of DoLP has the error approximate to 26 %. The reason is that the non-even of wood and different positions of image data and spectrometer data. Thus, the inversed polarization BRDF curve is presented to describe the real spectral features and textures better.

Materials for experiment are white tiles, plastic and aluminum board, black tiles, rubber, wooden board and green-painted board, white tiles and black rubber are used as white object and black body for inversion. Figure 4.9 shows the inversed polarization BRDF parameter curves at the position of $\theta_i = 24.72°$, $\theta_r = 42°$, $\varphi = 200°$. Curve f_{00} describes color and characteristics of material, the colors are presented to turn deeper in accordance with the curve order from top to bottom. The top three curves represent aluminum board, white tiles and white plastic, their f_{00} values are all over 0.2. And the f_{00} of wooden board changes in 0.05–0.2, while the value of green-painted board is close to 0.05. The values of black tiles and rubber are smaller than 0.05. Comparing with increasing of wooden board curve and decreasing of aluminum, white tiles, white plastic and black tiles curves are relatively flat. Peak appears in curve of green-painted board around 520 nm because 520 nm is the center wavelength of green light which is reflected a lot by green-painted board. All of the f_{00} values are positive. Compared with other materials, the value of aluminum is relatively large and the curve changes dramatically, which firstly drops and then rises in the range of 0.01–0.03. f_{20} is opposite to f_{00} that most of f_{20} values are negative. Curve f_{20} can also describe the value of f_{00}, the greater the value the sharper curve f_{20} changes. The DoLP curves of white tiles, aluminum board, white plastic and rubber are relatively flat. DoLP of rubber is around 0.35 and DoLP of white files and plastics are around 0.025. Aluminum has greater DoLP than other nonmetal materials with same color. The rest of materials has DoLP values among 0.025–0.35. The curve of black tiles and wooden board are decreasing, but the value of tiles is higher. The curve of green-pained board represents a trigonometric function shape. As for *Orient* curves, the white tiles and plastics curve change dramatically and the values of some bands are positive while others are negative.

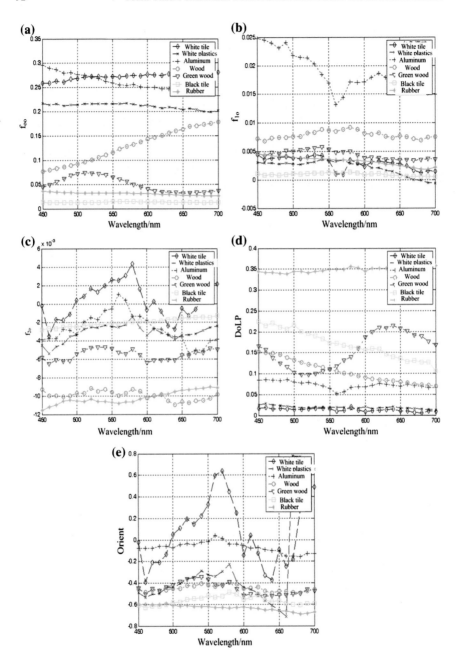

Fig. 4.9 Parameter curves of BRDF features. **a** f_{00} curves. **b** f_{10} curves. **c** f_{20} curves. **d** *DoLP* curves. **e** *Orient* curves

4.4 Polarization BRDF Model of Coated Targets

4.4.1 Intensity Component Model of Coated Targets

The roughness of the majority targets is between smooth surface and ideal diffuse surface. Reflection of incident light is a kind of directional diffuse reflection which is the characteristic of most rough surface. In this section, our studies are focus on the surface with directional diffuse reflection, the model of which is generally composed of directional reflection and diffuse reflection components.

(1) Exponential Model

The exponential BRDF model is a semi-empirical model which is conic, intuitive and simple. The model is a combination of a directional reflection function (exponential) and diffuses reflection function [2].

$$f_{00} = \frac{a \exp\left[-b(\theta_i - \theta_r)^2\right] + d \cos \theta_r}{\cos^c \theta_i \cos \theta_r} \tag{4.28}$$

where a, b, c, d are undetermined parameters. a determines the value of directional reflection; b determines valve width of specular reflection, the larger value of b, the sharper mirror peak, whereas the more smooth, c determines the change of scattering peak according to incident angle, d determines the value of diffuse reflection component.

(2) Micro-facet Model

Rough coated surface consists of an array of small patches with different normal directions so that the total scattering characteristics determined by every micro-facet [34]. The micro-facet model assumes that rough surface consists of an array of micro-facets which are the specular reflectors obeying Snell reflection theorem, its reflectivity is determined by Fresnel reflectivity $\rho(\beta)$ with local incident angle β. Every micro-facet is determined by a surface normal vector $\mathbf{n}$. Commonly used Gaussian distribution functions to describe the surface micro-facet height are as follows:

$$p(\theta) = \frac{1}{2\pi\sigma^2 \cos^3(\theta)} \exp\left[\frac{-\tan^2(\theta)}{2\sigma^2}\right] \tag{4.29}$$

As the formula is normalized,

$$\int d^2\omega p(\theta) = 1 \tag{4.30}$$

where the integration is defined in episphere. The slope of local surface is $\tan(\theta)$, the variance is σ^2. The height distribution of above assumptions is Gaussian that ensures $\theta < \pi/2$, from which the description of polarization BRDF is derived [34]:

$$f_{jl}(\theta_i, \theta_r, \phi) = \frac{1}{(2\pi)(4\sigma^2)(\cos\theta)^4} \frac{\exp\left[-\frac{\tan^2\theta}{2\sigma^2}\right]}{\cos(\theta_r)\cos(\theta_i)} \times m_{jl}(\theta_i, \theta_r, \phi) \qquad (4.31)$$

θ_i, θ_r, ϕ have the same physical definitions as those defined by polarization BRDF, θ is defined by

$$\cos(\theta) = \frac{\cos(\theta_i) + \cos(\theta_r)}{2\cos(\beta)} \qquad (4.32)$$

β is an auxiliary angle, which is the included angle of incidence light r_i and the normal direction n of small patch. Interactions between light and small patches satisfy the reflection law, so we have

$$\cos(\beta) = n \cdot r_i = n \cdot r_r \qquad (4.33)$$

$$\cos(2\beta) = \cos(\theta_i)\cos(\theta_r) + \sin(\theta_i)\sin(\theta_r)\cos(\phi) \qquad (4.34)$$

$m_{jl}(\theta_i, \theta_r, \phi)$ refers to Muller matrix. *Orient* parameters are sensitive to noise as it is the ratio of the third Stokes component and the second Stoles component, so it is not modeled in this text.

$$m_{00} = \frac{1}{2}(R_s + R_p) \qquad (4.35)$$

$$m_{10} = \frac{1}{2}\cos 2\eta_r (R_s - R_p) \qquad (4.36)$$

$$m_{20} = \frac{1}{2}\sin 2\eta_r (R_s - R_p) \qquad (4.37)$$

R_s and R_p are Fresnel reflection component equations which are parallel and perpendicular to the incident plane, η_r is the rotation angle of these two planes.

$$R_s = \frac{(A - \cos(\beta))^2 + B^2}{(A + \cos(\beta))^2 + B^2} \qquad (4.38)$$

$$R_p = R_s \left[\frac{(A - \sin(\beta)\tan(\beta))^2 + B^2}{(A + \sin(\beta)\tan(\beta))^2 + B^2}\right] \qquad (4.39)$$

$$A = \sqrt{\frac{\sqrt{C} + D}{2}}$$ (4.40)

$$B = \sqrt{\frac{\sqrt{C} - D}{2}}$$ (4.41)

$$C = 4n^2 k^2 + D^2$$ (4.42)

$$D = n^2 - k^2 - \sin^2(\beta)$$ (4.43)

In the sunlight, the mirco-facet based DoLP model is shown as follows:

$$
\begin{aligned}
DoLP(n, k) &= \frac{\sqrt{m_{10}^2 + m_{20}^2}}{m_{00}} \\
&= \frac{R_s - R_p}{R_s + R_p} = \frac{2A \sin^2(\beta) \cos(\beta)}{A^2 \cos^2(\beta) + \sin^4(\beta) + B^2 \cos^2(\beta)}
\end{aligned}
$$ (4.44)

From (4.31) and (4.44), there are many physical quantities which are difficult to determine in intensity components, such as surface roughness σ, refractive index n, k so they are difficult to model.

(3) Simulation Results and Analysis

The images of green-painted aluminum board are collected in 9 different incident positions and receiving positions, there are 33 spectral bands. The data of seven positions are used in parameter inversion, while the data of rest two positions are used in comparison with model predictive data. Table 4.1 and Fig. 4.10 show the comparison of predictive values and experimental values. From the results we can know that average error of f_{00} in two positions are 10.6 and 8.3%. Exponential model is a relatively simple model to describe f_{00} component of coated targets.

4.4.2 DoLP Model of Coated Targets

The relationship between DoLP and phase angle is simulated by the polynomials models. The DoLP of seven positions are used in parameter inversion, while the DoLP of rest two positions the others are used in comparison with predictive data. Table 4.2 and Fig. 4.11 show the comparison of them.

The average errors of DoLP in two positions are 16.2 and 18.4 %, a quintic polynomial model can describe DoLP of both ground materials and coated materials.

Table 4.1 Predicted and experimental values of f_{00} component in green-painted aluminum board

(a) $\theta_i = 52.11°, \theta_r = 40°, \phi = 30°$

Wavelength (nm)	Experimental value (sr^{-1})	Predictive value (sr^{-1})	Error (%)
450	0.021807	0.01881	13.743
500	0.045977	0.03819	16.937
550	0.032747	0.028543	12.838
600	0.016768	0.014321	14.593
650	0.015029	0.014823	1.3692
700	0.019137	0.0191	0.19487

(b) $\theta_i = 55.97°, \theta_r = 35°, \phi = 90°$

Wavelength (nm)	Experimental values (sr^{-1})	Predictive values (sr^{-1})	Error (%)
450	0.016986	0.015455	9.0118
500	0.037242	0.033406	10.301
550	0.027034	0.025333	6.2923
600	0.012075	0.012575	4.1481
650	0.010057	0.010608	5.4814
700	0.013284	0.013569	2.1462

(c) Multi-band f_{00} parameters

Wavelength (nm)	a	b	c	d
450	3.1876	838.32	−2.1117	0.05266
500	2.6601	760.47	−1.4385	0.077
550	1.0095	628.96	−1.2824	0.05333
600	0.15999	303.24	−1.3966	0.02829
650	0.044117	53.475	−1.0896	0.01993

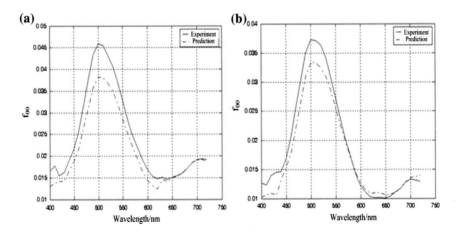

Fig. 4.10 Spectral curve of f_{00} experimental values and predictive values. **a** $\theta_i = 55.11°, \theta_r = 40°, \phi = 30°$. **b** $\theta_i = 55.97°, \theta_r = 35°, \phi = 90°$

Table 4.2 Comparison of predictive values and experimental values of green-painted aluminum board DoLP components

(a) $\theta_i = 47.39°, \theta_r = 15°, \phi = 0°$

Wavelength (nm)	Experimental value (sr^{-1})	Predictive value (sr^{-1})	Error (%)
450	0.34169	0.28933	15.324
500	0.30546	0.25246	17.351
550	0.31445	0.27281	13.243
600	0.32569	0.2715	16.64
650	0.30447	0.25394	16.596
700	0.3489	0.30545	12.453

(b) $\theta_i = 55.11°, \theta_r = 40°, \phi = 30°$

Wavelength (nm)	Experimental value (sr^{-1})	Predictive value (sr^{-1})	Error (%)
450	0.085139	0.10177	19.538
500	0.087095	0.10354	18.881
550	0.093426	0.10524	12.643
600	0.090594	0.10827	19.515
650	0.074662	0.09146	22.496
700	0.12197	0.13522	10.867

(c) Multi-band DoLP parameters

Wavelength (nm)	P_0	P_1	P_2	P_3	P_4	P_5
450	1.561	−26.179	113.25	−196.42	148.57	−40.704
500	1.2512	−20.867	90.83	−158.56	120.9	−33.443
550	1.4333	−23.661	102.44	−178.38	135.81	−37.554
600	1.6557	−24.312	101.52	−173.54	130.52	−35.741
650	1.6508	−22.914	93.177	−156.42	116.05	−31.424
700	1.5295	−22.24	92.634	−157.22	117.45	−31.994

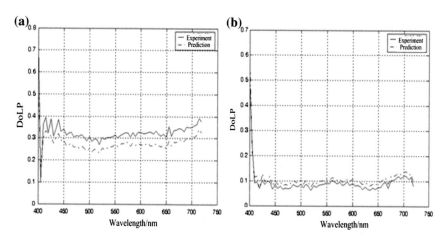

Fig. 4.11 Spectral curve of DoLP experimental values and predictive values. **a** $\theta_i = 47.39°, \theta_r = 15°, \phi = 0°$. **b** $\theta_i = 55.11°, \theta_r = 40°, \phi = 30°$

4.4.3   Relationships of Polarization BRDF, Azimuth Angle, and Wavelength

This section analyzes the green-coated targets changing with azimuth angle and wavelength. The target was cleaned before the experiment to ensure no polluting particles. Figure 4.12 shows the relationship of polarization BRDF f_{00}, detecting angle and wavelength. The range of detecting angle and wavelength are respectively 5°–40° and 400–720 nm. From Fig. 4.12 we know that f_{00} changes significantly in the wavelength of 400–600 nm and arrives the maximum value at 530 nm, which is exactly the green light band. f_{00} shows the reflection of incidence light on the surface of wooden board, which is proportional to the reflectivity. In the range of 600–720 nm, f_{00} is essentially a constant. When the observation angle changes from 5° to 20°, f_{00} does not change significantly and proves to have stronger Lambertian characteristics. However, non-Lambertian characteristics are more obvious when the observation angle is more than 20°, f_{00} increases dramatically when observation angle changes from 20° to 40°. From Fig. 4.12 we know that DoLP of green-painted coated board is relatively higher, whose maximum value is

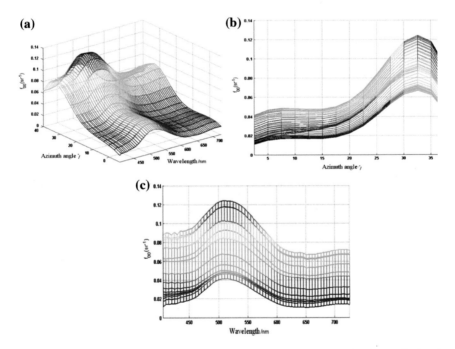

Fig. 4.12 The f_{00} curve of green-coated board in different wavelength with the change of azimuth

0.6. The curve which shows the DoLP changing with wavelength is opposite to the curve which shows f_{00} changing with wavelength. DoLP shows obvious minimum value when wavelength is near 530 nm, which reflects that DoLP is changed opposite to the reflectivity. The differences between larger and smaller azimuth angle are more obvious, which is due to the experimental value is small and affected more by noises. When azimuth angle is larger, DoLP changes as a single-peak curve that monotone increases and then decreases rapidly.

Comparing two kind of target with different materials but the same coated color in Figs. 4.13 and 4.14, we know that polarization BRDF parameter f_{00} of these two targets are nearly the same, as well as its relationship with wavelength and azimuth angle. Apart from this, DoLP also shows the similar characteristic according to Figs. 4.14 and 4.15, which is because that only a small part of light can reach target surface and have interaction with the internal part, then arrives detector. The received reflecting energy is mainly originated from the single or multi scattering on surface. So polarization information of target structure is mostly detected and different materials with the same coated color have similar structures.

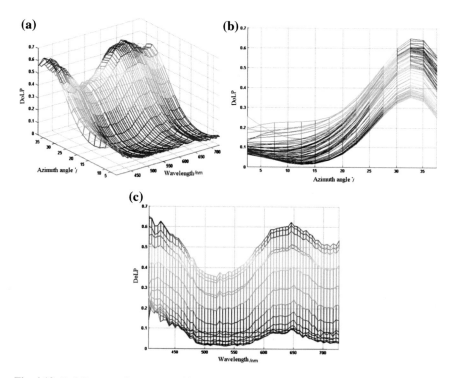

Fig. 4.13 DoLP curve of green-coated board in different wavelengths with the change of azimuth

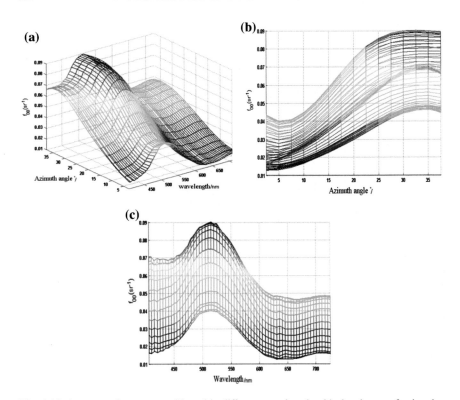

Fig. 4.14 f_{00} curve of green-coated board in different wavelength with the change of azimuth

4.5 Polarization BRDF Model of the Background

Comparing with P-BRDF model, the background models are empirical without solid physical foundations, whose purpose is providing a method to quantify the material P-BRDF. The obtained first column (f_{00}, f_{10}, and f_{20}) of P-BRDF matrix cannot completely describe incident polarization radiation. Thus, P-BRDF model of background includes the following three aspects:

- Intensity component (f_{00}) model
- DoLP model;
- Polarization Orient model

Parameters of these models are incident zenith angle, receiving zenith angle and azimuth angle, a form which we can obtain f_{10} and f_{20}.

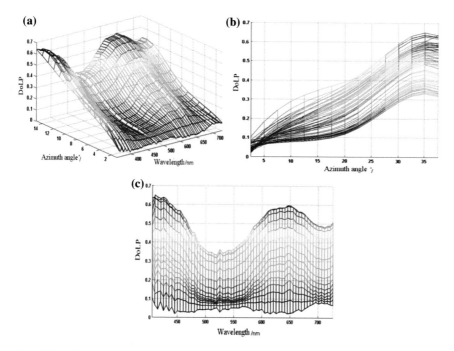

Fig. 4.15 DoLP curve of green-coated board in different wavelengths with the change of azimuth

$$
f_{10} = \frac{DoLP \times f_{00}}{\sqrt{1 + \tan^2(2 \times AoP)}}
\begin{cases}
+f_{10} & -\frac{\pi}{4} \le Orient \le \frac{\pi}{4} \\
-f_{10} & -\frac{\pi}{2} \le Orient \le \frac{\pi}{2} \ or \ \frac{\pi}{4} \le Orient \le \frac{\pi}{2}
\end{cases}
\tag{4.45}
$$

$$
f_{20} = \frac{DoLP \times f_{00} \times \tan(2 \times Orient)}{\sqrt{1 + \tan^2(2 \times Orient)}}
\begin{cases}
+f_{20} & 0 \le Orient \le \frac{\pi}{2} \\
-f_{20} & -\frac{\pi}{2} \le Orient \le 0
\end{cases}
\tag{4.46}
$$

4.5.1   Intensity Component Model of the Background

(A)   Description

(1)   Rojean Model

Rojean model is a kind of semi-empirical, kernel BRDF model that describes intensity components and shows great benefit to description of material BRDF. Rojean model consists of three kernels by weighted linear combination:

- Isotropic scattering function
- Geometrical scattering function (Consider the shadow and salient on the surface of background)

- Volume scattering function [The random mirco-faces scattering effect based on radiative transfer which is show in (4.48)]

$$f_{00}(\theta_i, \theta_r, \phi, \lambda) = k_0 + k_1 f_1(\theta_i, \theta_r, \phi, \lambda) + k_2 f_2(\theta_i, \theta_r, \phi, \lambda) \qquad (4.47)$$

Function f_1 is obtained considering the geometrical salient, function f_2 is converted through simple radiative model.

$$f_1(\theta_i, \theta_r, \phi, \lambda) = \frac{1}{2\pi}[(\pi - \phi)\cos\phi + \sin\phi]\tan\theta_i \tan\theta_r$$
$$- \frac{1}{\pi}(\tan\theta_i + \tan\theta_r + \sqrt{\tan^2\theta_i + \tan^2\theta_r - 2\tan\theta_i \tan\theta_r \cos\phi})$$

$$(4.48)$$

$$f_2(\theta_i, \theta_r, \phi, \lambda) = \frac{2}{3\pi}\frac{(\pi - 2\xi)\cos\xi + 2\sin\xi}{\cos\theta_i + \cos\theta_r} - \frac{1}{3} \qquad (4.49)$$

Phase angle ξ can be represented as:

$$\cos\xi = \cos\theta_i \cos\theta_r + \sin\theta_i \sin\theta_r \cos\phi \qquad (4.50)$$

where $0 \le \phi \le \pi$, $0 \le \theta_i, \theta_r \le \pi/2$. The azimuth angle of backscattering is $\phi = 0$. Parameters k_0, k_1, and k_2 are respectively the weights of corresponding scattering function.

(2) Waltthall model

The Waltthall model is an empirical model with reciprocity which is widely applied to simulating soil BRDF.

$$f_{00}(\theta_i, \theta_r, \phi, \lambda) = p_0(\theta_i^2 + \theta_r^2) + p_1\theta_i^2\theta_r^2 + p_2\theta_i\theta_r \cos\phi + p_3 \qquad (4.51)$$

Here p_0, p_1, p_2 and p_3 are four linear parameters. Walthall cannot simulate some significant surface characteristics such as groove structure and the peak of hot spot.

(B) Inversion

Linear Rojean model and Walthall model can be inversed through matrix transformation so that it is fast and suitable for mixed pixels. Here Rojean model is taken as an example to derive the inversion of multi-band BRDF model parameters. Least Square method is applied to inverse parameters k_0, k_1, and k_2. More than three groups of experimental results are used to solve overdetermined equations.

Assuming that the material is measured for n times in m different bands, the experimental data is shown:

$$\mathbf{F_{00}} = \mathbf{K}\mathbf{F} \tag{4.52}$$

$\mathbf{F_{00}}$ is a measured matrix with the size of $m \times n$, while $\mathbf{K} = [\mathbf{K_0}, \mathbf{K_1}, \mathbf{K_2}]$ is the Rojean parameter matrix with the size of $m \times 3$. $\mathbf{F}$ denotes the scattering function matrix with the size of $3 \times n$ and the ith row represents the column vector $[1 \quad f_1^{(i)} \quad f_2^{(i)}]^T$ of the ith measurement. $\mathbf{F} = \begin{bmatrix} 1 & \cdots & 1 \\ f_1^{(1)} & \cdots & f_1^{(n)} \\ f_2^{(1)} & \cdots & f_2^{(n)} \end{bmatrix}$.

So the Least Square solution of $\mathbf{K}$ is

$$\mathbf{K} = \mathbf{F_{00}}\mathbf{F}^{\#} \tag{4.53}$$

where $\mathbf{F}^{\#}$ is the pseudo inverse of $\mathbf{F}$, $\mathbf{F}^{\#} = \mathbf{F}^{T}(\mathbf{F}\mathbf{F}^{T})^{-1}$. So the model parameters k_0, k_1, and k_2 can be determined and the values of f_{00} of any incident and receiving directions in multi-band data can also be predicted through these parameters. Inversion of Walthall model is similar to Rojean so it is omitted in this section.

(C)   Simulation and Analysis
(1)   Rojean Model

Data measured on grass is used to verify Rojean model. The background is measured in 33 bands (400–720 nm with a step of 10 nm) at 10 different incident and receiving positions, seven of which are used to inverse and the others are used to compare with the predictive data as shown in Table 4.3 and Fig. 4.16a–c in Fig. 4.16 are corresponding to (a), (b), and (c) in Table 4.3.

From the result we know that the average errors of grass f_{00} at 3 positions are less than 9 % while the error in 400–450 nm is relatively greater, which is because of the weak light and strong noise passed LCTF. Rojean model is a steady BRDF model that can accurately describe the f_{00} component of grass.

(2)   Walthall model

In the experiments, the soil is dried and grinded as mixture of tiny particles and powder. The background is measured in 33 bands at 11 different incident and receiving positions, eight of which are used to inverse and the others are used to compare with the predictive data shown in Table 4.4 and Fig. 4.17. From the result we know that the average errors of soil f_{00} components at 3 positions are less than 5 %. Walthall is an accurate model that describes the f_{00} component of soil precisely.

Table 4.3 Comparison of predictive values and experimental values of f_{00} component of grass

(a) $\theta_i = 60.17°, \theta_r = 5°, \phi = 60°$

Wavelength (nm)	Experimental values (sr^{-1})	Predictive values (sr^{-1})	Error (%)
450	0.006867	0.0059602	13.205
500	0.0082193	0.0079627	3.1221
550	0.026693	0.027035	1.2839
600	0.017637	0.017236	2.275
650	0.01134	0.010205	10.007
700	0.032189	0.032006	0.56968

(b) $\theta_i = 56.41°, \theta_r = 30°, \phi = 0°$

Wavelength (nm)	Experimental values (sr^{-1})	Predictive values (sr^{-1})	Error (%)
450	0.01199	0.013451	12.187
500	0.014146	0.014005	0.99856
550	0.031892	0.035738	12.059
600	0.023697	0.025754	8.8627
650	0.016973	0.017584	3.6016
700	0.037706	0.040663	7.8428

(c) $\theta_i = 62.23°, \theta_r = 15°, \phi = 90°$

Wavelength (nm)	Experimental values (sr^{-1})	Predictive values (sr^{-1})	Error (%)
450	0.0073576	0.0062186	15.48
500	0.0089033	0.0084451	5.1475
550	0.027601	0.030079	8.9804
600	0.018181	0.018588	2.237
650	0.011225	0.010594	5.6189
700	0.031852	0.034585	8.5787

(d) Multi-band f_{00} parameters

Wavelength (nm)	k_0	k_1	k_2
450	1.6909	−0.34226	24.302
500	1.7529	−0.85658	22.675
550	2.2502	−6.2122	59.032
600	2.8826	−2.6311	39.509
650	2.7309	−0.62272	25.451
700	4.8319	−5.2278	53.682

4.5.2 DoLP Model of the Background

DoLP is closely related to phase angle ζ. We have found that quintic polynomial can perfectly fit in the experimental data, the quintic polynomial is shown as follows:

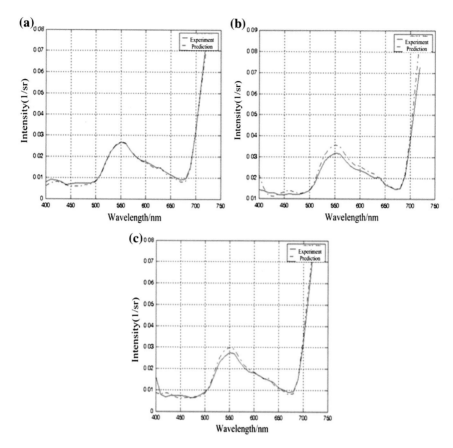

Fig. 4.16 Spectral curves of predictive values and experimental values of f_{00} component on grass.
a $\theta_i = 60.17°, \theta_r = 5°, \phi = 60°$, **b** $\theta_i = 56.41°, \theta_r = 30°, \phi = 0°$, **c** $\theta_i = 62.23°, \theta_r = 15°$,
$\phi = 90°$

$$DoLP = p_0 + p_1 \xi + p_2 \xi^2 + p_3 \xi^3 + p_4 \xi^4 + p_5 \xi^5 \qquad (4.54)$$

Figure 4.18 shows the data of soil at 450 and 650 nm and the relationship between data and the quintic polynomial which can describe the reaction of DoLP and phase angle. In the experiment, phase angles are gathered in the range of $0-\pi/2$, while it lacks verification and analysis in the range of $\pi/2-\pi$. So the following conclusions can be draw:

(1) DoLP decreases gradually when phase angle decreases from $\pi/6$ to 0 until arrives at a certain critical angle, after which DoLP increases rapidly.
(2) DoLP increases gradually when phase angle increases from $\pi/3$ to $\pi/2$ until arrives at a certain critical angle, after which DoLP decreases.

Table 4.4 Comparison of predictive values and experimental values of f_{00} component of soil

(a) $\theta_i = 41.00°, \theta_r = 45°, \phi = 45°$

Wavelength (nm)	Experimental values (sr⁻¹)	Predictive values (sr⁻¹)	Error (%)
450	0.06291	0.064002	1.7358
500	0.07295	0.074776	2.5025
550	0.08965	0.092913	3.6399
600	0.10661	0.11174	4.8068
650	0.1147	0.12098	5.4777
700	0.12086	0.12768	5.6409

(b) $\theta_i = 34.72°, \theta_r = 15°, \phi = 90°$

Wavelength (nm)	Experimental values (sr⁻¹)	Predictive values (sr⁻¹)	Error (%)
450	0.047696	0.047892	0.40995
500	0.055712	0.056055	0.61522
550	0.06947	0.07038	1.3102
600	0.08336	0.085	1.9673
650	0.090443	0.092078	1.8078
700	0.095332	0.098019	2.8184

(c) $\theta_i = 44.47°, \theta_r = 10°, \phi = 180°$

Wavelength (nm)	Experimental values (sr⁻¹)	Predictive values (sr⁻¹)	Error (%)
450	0.035961	0.038509	7.0865
500	0.042663	0.044795	4.9987
550	0.053397	0.055864	4.6215
600	0.064394	0.066962	3.9877
650	0.069562	0.07204	3.5616
700	0.074437	0.076774	3.1398

(d) Multi-band parameters of f_{00} model

Wavelength (nm)	p_0	p_1	p_2	p_3
450	−2.30	6.60	3.25	5.63
500	−2.88	8.24	3.71	6.65
550	−3.88	10.79	4.56	8.46
600	−4.97	13.67	5.40	10.32
650	−5.67	15.57	5.77	11.29
700	−6.06	16.38	6.05	12.03

DoLP at nine positions are used to inverse parameters in the quintic polynomial while others are used for comparison and analysis. Table 4.5 and Fig. 4.19 show the comparison of predictive values and experimental values.

From the simulations, we know that the average errors of DoLP at these two positions are 5.6 and 9.4 %, respectively. Although a quintic polynomial model is obtained through the observation and analysis, it accurately describes the variation of DoLP component.

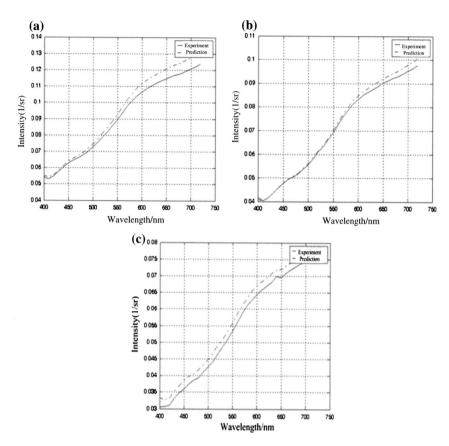

Fig. 4.17 Spectral curves of predictive values and experimental values of f_{00} component on grass. **a** $\theta_i = 41.00°, \theta_r = 45°, \phi = 45°$, **b** $\theta_i = 34.72°, \theta_r = 15°, \phi = 90°$, **c** $\theta_i = 44.47°, \theta_r = 10°$, $\phi = 180°$

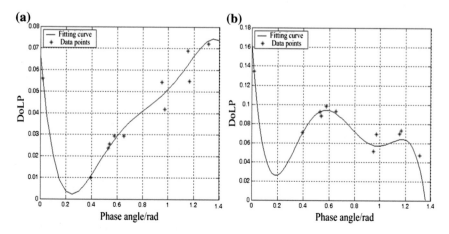

Fig. 4.18 Fitting curves of soil data. **a** 450 nm, **b** 650 nm

Table 4.5 Comparison of predictive values and experimental values of DoLP component of soil

(a) $\theta_i = 41.00°, \theta_r = 45°, \phi = 45°$

Wavelength (nm)	Experimental values (sr^{-1})	Predictive value (sr^{-1})	Error (%)
450	0.023782	0.024956	4.9377
500	0.055232	0.054211	1.8484
550	0.063887	0.063091	1.2449
600	0.081448	0.080069	1.6933
650	0.092201	0.092182	0.02072
700	0.14151	0.14004	1.0404

(b) $\theta_i = 34.72°, \theta_r = 15°, \phi = 90°$

Wavelength (nm)	Experimental values (sr^{-1})	Predictive value (sr^{-1})	Error (%)
450	0.029454	0.035059	19.03
500	0.058076	0.057084	1.7086
550	0.063649	0.062134	2.381
600	0.081795	0.078035	4.5964
650	0.092829	0.089896	3.1602
700	0.14337	0.13917	2.9254

(c) Multi-band parameters of the DoLP model

Wavelength (nm)	p_0	p_1	p_2	p_3	p_4	p_5
450	0.067275	−0.73432	2.6491	−3.8264	2.5062	−0.61125
500	0.11842	−1.9479	8.6827	−14.953	11.185	−3.0479
550	0.13044	−1.9528	8.8212	−15.45	11.723	−3.2352
600	0.15008	−1.8419	8.2348	−14.371	10.865	−2.9873
650	0.16155	−1.7822	7.9385	−13.821	10.419	−2.855
700	0.21108	−1.7531	7.6719	−13.147	9.7676	−2.6416

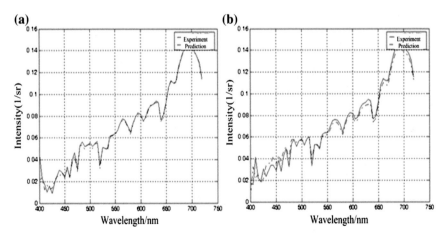

Fig. 4.19 Predicted and experimental spectral curves of DoLP component. **a** $\theta_i = 41.00°, \theta_r = 45°, \phi = 45°$, **b** $\theta_i = 34.72°, \theta_r = 15°, \phi = 90°$

References

1. Nicodemus FE, HSIA JJ, Ginsberg IW, Limperis T (1977) Geometrical considerations and nomenclature for reflectance. National Bureau of Standards, Washington, D. C., pp 1–7
2. Xu X (2005) Remote sensing physics. Beijing University Press, Beijing
3. Gao J, Ma J, Jin Z (1997) Multi-angle and Multi-spectral remote sensing technique. Spacecr Recovery Remote Sens 18(2):26–30
4. Liao Y (2003) Polarization optics. Science Press, Beijing
5. Hahlweg C, Rothe H (2005) Design of a full-hemispherical spectro-radiometer with high dynamic range for characterization of surface properties using multi-spectral BRDF data form VIS to NIR. Proc SPIE, p 5965
6. Hahlweg C, Rothe H (2007) Spectroscopic imaging from 400 nm to 1800 nm with liquid crystal tunable filters. Proc SPIE, p 6503
7. Davis MH (1990) A CCD based bidirectional spectral reflectance field instrument, Master's Thesis, Rochester Institute of Technology
8. Chen C (2010) Light scattering and radiation characteristic of target and environment and its application. School of Automation, Northwestern Polytechnical University, Xi'an
9. Shell JR (2005) Polarimetric remote wensing in the visible to near infrared, Ph.D. Dissertation, Rochester Institute of Technology
10. Sandmeier SR and Itten KI (1999) A field goniometer system (FIGOS) for acquisition of hyperspectral BRDF data. IEEE Trans Geosci Remote Sens 37(2):978–986
11. Sandmeier SR (2000) Acquisition of bidirectional reflectance factor data with field goniometers. Remote Sens Environ 73(3):257–269
12. Shell JR, Schott JR (2005) A polarized clutter measurement technique based on the governing equation for polarimetric remote sensing in the visible to near infrared. Proc SPIE 5811:34–45
13. Ward GJ (1992) Measuring and modeling anisotropic reflection. Comput Graph 26 (2):265–272
14. Torrance KE, Sparrow EM (1966) Off-specular peaks in the directional distribution of reflected thermal radiation. J Heat Transfer-Trans ASME, 223–230
15. Maxwell JR, Beard J, Weimer S, Ladd D, Ladd S (1973) Bidirectional reflectance model validation and utilization, Technical Report AFAL-TR-73-303, Environmental Research Institute of Michigan
16. Conant JA, Iannarilli FJ Jr (2002) Development of a combined bidirectional reflectance and directional emittance model for polarized modeling. Proc SPIE 4481:206–215
17. Roujean JL, Latoy M, Deschamps PY (1992) A bidirectional reflectance model of the earth's surface for the correction of remote sensing data. J Geophys Res 97:20455–20468
18. Hapke BW (1981) Bidirectional reflectance spectroscopy theory. J Geophys Res 86:3039–3054
19. Walthall CL, Norman JM, Welles JM, Campbell G, Blad BL (1985) Simple equation to approximate the bidirectional reflectance from vegetation canopies and bare soil surfaces. Appl Opt 24:383–387
20. Priest RG, Germer TA (2000) Polarimetric BRDF in the microfacet model: theory and measurements. In: Proceedings of meeting of the military sensing symposia specialty group on passive sensor, vol 1, pp 169–181
21. Chung AJ, Deligianni F, Shah P, Wells A, Wells GZ (2006) Patient-specific bronchoscopy visualization through BRDF estimation and disocclusion correction. IEEE Trans Med Imaging 25(4):503–513
22. Li H, Torrance KE (2005) An experimental study of the correlation between surface roughness and light scattering for rough metallic surfaces. Proc SPIE, p 5878
23. Breon F-M, Tanre D, Lecomte P, Herman M (1995) Polarized reflectance of bare soils and vegetation: measurements and modals. IEEE Trans Geosci Remote Sens 33(2):487–498

24. Fujikake H, Takizawa K, Aida T, Kikuchi H, Fujii T, Kawakita M (1998) Electrically-controllable liquid crystal polarizing filter for eliminating reflected light. Opt Rev 5(2):93–98
25. Hapke B (1981) Bidirectional reflectance spectroscopy: 1. Theory. J Geophys Res Solid Earth 86(B4):3039–3054
26. Ashikhmin M, Shirley P (2000) An anisotropic phong BRDF model. J Graph Tools 5 (2):25–32
27. Walthall C, Roujean JR, Morisette J (2000) Field and landscape BRDF optical wavelength measurements: Experience, techniques and the future. Remote Sens Rev 18(2–4):503–531
28. Otterman J, Brakke TW, Susskind J (1997) A model for inferring canopy and underlying soil temperatures from multi-directional measurements. Bound-Layer Meteorol 61(1–2):81–97
29. Frank SW, Suttles JT (1986) Reflection and emission models for deserts derived from Nimbus-7 ERB scanner measurements. J Climate Appl Meteorol 25(2):196–202
30. Torrance KE, Sparrow EM (1967) Theory for off-specular reflection from roughened surfaces. J OSA 57(9):1105–1112
31. Cook RL, Torrance KE (1982) A reflectance model for computer graphics. ACM Trans Graph 1(1):7–24
32. Maxwell JR, Beard J, Weiner S, Ladd D, Ladd S (1973) Bidirectional reflectance model validation and utilization, Technical Report AFAL-TR-73-303, Wright Patterson AFB
33. Sanford BP, Robertson DC (1985) Infrared reflectance properties of aircraft paints. Proc IRIS, pp 111–127
34. Priest RG, Meier SR (2002) Polarimetric microfacet scattering theory with applications to absorptive and reflective surfaces. Opt Eng 41(5):988–993
35. Chen C, Zhao Y, Pan Q (2009) Models validation and analysis of spectropolarimatric BRDF for soil. J Optoelectron Laser 20(3):369–373
36. Chao C, Zhao Y, Luo L, Liu D, Pan Q (2009) Robust materials classification based on multispectral polarimetric BRDF imagery. Proc SPIE
37. Chen C, Zhao Y, Pan Q (2010) Classification based on Spectropolarimetric BRDF. Acta Photonica Sinica 39(6):1026–1033
38. Yang T, Zhao Y, Pan Q (2008) A new measurement method for polarized spectral bidirectional reflection. Acta Photonica Sinica 37(12):2520–2524
39. Hess M, Priest R (1999) Comparison of polarization bidirectional reflectance distribution function (BRDF) models. IEEE Aerosp Conf Proc 4:95–102
40. Georgiev GT, Butler JJ (2004) The effect of incident light polarization on spectral BRDF measurements. Proc SPIE 5570:492–502

Chapter 5
Object Detection with Multi-band Polarization Imaging

Multi-band polarization imaging provides complementary information to conventional imaging sensors, such as material components and surface characteristics. An image data in multi-band polarization imaging makes an array of 3D data cubes. Each pixel corresponds to a certain point in a high dimensional space where a statistical analysis of the data leads to object detection. There are two aspects in object detection using multi-band polarization imaging. From a theoretical background of multivariate statistical analysis, multi-band polarization image data can be considered as common multivariate data to express statistical differences between the target object and the background. Physical properties of multi-band polarization imagery can be replaced by statistical differences to make the traditional optical remote sensing processing methods useful. Multi-band polarization imagery contains spatial and spectral information as well as polarization information, which is useful for object detection in challenging conditions. Physical laws of reflection, scattering, and radiation of polarization and spectrum of the light such as Fresnel law and Malus law are used to identify the object. Object detection with multi-band polarization imagery can be applied in signal processing, information fusion, and multivariate data analysis. This chapter describes some of latest object detection methods based on information fusion, along with examples in the foundation of the basic detection theory and mathematical methods.

5.1 Multi-band Polarization Image Analysis

According to the imaging chain theory [1], object detection capability can be improved by adding more information from the scene. The object detection performance can be improved by fusing spatial, spectral, and polarization information of the same object in the scene. Multi-band polarization image data provide spatial, spectral, and polarization information, which can be represented by 7 independent variables: spatial coordinates (x, y), wavelength (λ), and four polarization variables

© National Defense Industry Press, Beijing and Springer-Verlag Berlin Heidelberg 2016   111
Y. Zhao et al., *Multi-band Polarization Imaging and Applications*,
Advances in Computer Vision and Pattern Recognition,
DOI 10.1007/978-3-662-49373-1_5

(S_0, S_1, S_2, S_3). Multi-band polarization imaging utilizes all the information comprehensively to improve object detection performance. Extracting the information from multi-band polarization imagery is the premise of precise detection due to its high dimension and multi-modes. Figure 5.1 describes the process that the real space is projected onto the 7-dimensional polarization space in multi-band polarization imaging.

In multi-band polarization imaging, an observed data in K spectral bands can be described as a point denoted as VS^K in K-dimensional vector space $\begin{bmatrix} S_{0,\lambda} & S_{1,\lambda} & S_{2,\lambda} \\ DoLP_\lambda & Orient_\lambda \end{bmatrix}$ $\lambda = 1, 2, \ldots, K$. Every band is designated as an axis and all of these axes are mutually orthogonal [2], so the multi-band polarization properties of every pixel can be described by $\mathbf{X} = \begin{bmatrix} sp_{S_0} & sp_{S_1} & sp_{S_2} & sp_{DoLP} & sp_{Orient} \end{bmatrix}$, where $sp_p = \begin{bmatrix} sp_{p_1} & sp_{p_2} & \cdots & sp_{p_\lambda} & \cdots & sp_{p_K} \end{bmatrix}$ $p \in \{S_0, S_1, S_2, DoLP, Orient\}$. A column corresponds to a K-dimensional vector, and every element denotes polarized reflectance of each spectral band, or polarized spectrum. In multi-band polarization imaging, the similarity of polarized spectrum X_1 and X_2 can be measured by the Euclidean distance or Spectral Angle Mapper (SAM)

$$\|\mathbf{X}_1 - \mathbf{X}_2\| = \sqrt{\sum_{\lambda=1}^{K} (sp1_{\lambda S_0} - sp2_{\lambda S_0})^2 + (sp1_{\lambda S_1} - sp2_{\lambda S_1})^2 + (sp1_{\lambda S_2} - sp2_{\lambda S_2})^2 + (sp1_{\lambda DoLP} - sp2_{\lambda DoLP})^2 + (sp1_{\lambda Orient} - sp2_{\lambda Orient})^2}$$

$$(5.1)$$

$$\theta(\mathbf{X}_1, \mathbf{X}_2) = \sum_{p \in \{S_0, S_1, S_2, DoLP, Orient\}} w_p \cos^{-1}\left(\frac{\langle sp1_p, sp2_p \rangle}{\|sp1_p\| \|sp2_p\|}\right) \qquad (5.2)$$

where $\langle \rangle$ represents dot product and ω_p is a weight.

5.2 Object Detection

5.2.1 Polarization Mixture Model

Due to a low spatial resolution in multi-band spectral imaging, a single pixel is composed of multiple materials. Therefore, the observed data is compositely generated from spectral features of different materials. A mixture model of polarized spectrum describes the quantitative relationship between observation values and different spectrum features. From Chap. 4, observed data of polarization images is affected by the factors such as illumination angles, atmospheric environments, second irradiation, and material compositions. However, only material compositions are considered in the polarized spectrum mixture model [3].

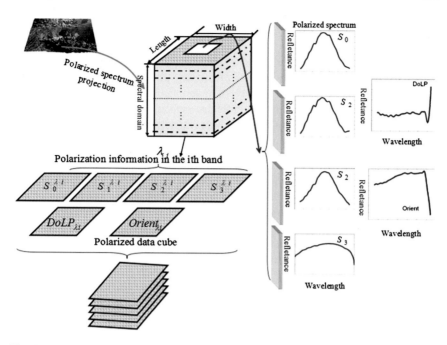

Fig. 5.1 Projection of multi-band polarization image data

Definition 5.1 A linear polarized spectrum mixture model is given by:

$$\mathbf{x} = \sum_{i=1}^{M} \alpha_i s_i + \mathbf{w} = S\alpha + \mathbf{w} \qquad (5.3)$$

where x denotes a pixel in polarization image, s_i a polarized spectrum vector of the i-th material or endmember; α_i is the fractional abundance of the i-th endmember, and w gives an observation error.

Definition 5.2 A nonlinear polarized spectrum mixture model is given by [3]:

$$\mathbf{x} = f(\mathbf{S}) + \mathbf{w} \qquad (5.4)$$

where $\mathbf{S} = [\mathbf{s}_1, \mathbf{s}_2, \ldots, \mathbf{s}_M]$ represents polarized spectrum features. A nonlinear function $f(\cdot)$ generally has multiple forms in accordance with different nonlinear models.

5.2.2 Multivariate Gaussian Distribution

Multivariate normal distribution is the fundamental of multivariate data analysis and many multivariate statistical signal processing theories. Multivariate normal assumption can simplify theoretical derivation and analysis. The data distributions referred in many realistic multivariate problem, especially the problems under the conditions of huge samples, are always approximately normal [4, 5]. Therefore, multivariate normal distribution can be significant to multivariate signal and data analysis.

Definition 5.3 There is a probability distribution family $\mathcal{P}$ in a measureable space $(\mathcal{H}, \mathcal{B})$. And $(\mathcal{H}, \mathcal{B}, \mathcal{P})$ represents the statistical model. If P depends only on parameter θ, which is denoted as $\mathcal{P} = \{\mathcal{P}_\theta \neg \theta \in \Theta\}$ (Θ is the parametric space), there will be a parametric model. Generally, a typical parametric model consists of the multivariate normal distribution family and the corresponding measureable space.

Definition 5.4 Suppose that $N_p(\mu, \Sigma)$ is a p-dimensional constant vector, Σ denotes a $p \times p$ positive definite matrix. The probability density function of random vector $\mathbf{x} = (x_1, x_2, \ldots, x_p)^T$ in the measureable space $(R^p, \mathcal{B}_{R^p})$ is given by:

$$f_{X_1, X_2, \ldots, X_p}(x_1, x_2, \ldots, x_p) = \frac{1}{(2\pi)^{p/2} |\Sigma|^{1/2}} \exp\left\{ -\frac{1}{2} (\mathbf{x} - \mu)^T \Sigma^{-1} (\mathbf{x} - \mu) \right\} \quad (5.5)$$

Definition 5.5 For a p-dimensional normal random vector $\mathbf{y} = (y_1, y_2, \ldots, y_p)^T$ with the Gaussian distribution function $N_p(\mu, \Sigma)$ and $\text{Rank}(\Sigma) = r$, the mean $E(y) = \mu$ and the covariance matrix Σ,

$$\mathbf{y} = (y_1, y_2, \ldots, y_p)^T \quad (5.6)$$

Definition 5.6 Suppose that μ denotes a p-dimensional vector, $\mathbf{B}$ is a $p \times r$ matrix with rank r, $\mathbf{U} \sim N_r(0, \mathbf{I}_r)$. If the distributions of a p-dimensional random vector $\mathbf{z} = (z_1, z_2, \ldots, z_p)^T$ and $\mu + \mathbf{BU}$ are the same, then $\mathbf{z}$ has a distribution function $N_p(\mu, \Sigma)$, where $\Sigma = \mathbf{BB}^T$ and r is the rank.

5.2.3 Maximum Likelihood Estimation

Maximum likelihood estimation has been one of the most significant and widely used parameter estimation method [4]. The function $p_{\mathbf{x}}(\mathbf{x}; \theta)$ shows how probability changes with sample $\mathbf{x}$ when θ is known. On the other hand, $p_{\mathbf{x}}(\mathbf{x}; \theta)$ shows how probability changes with different θ when sample $\mathbf{x}$ is known.

Definition 5.7 Assuming that $\boldsymbol{\theta} \in \Theta$ is a family of joint probability density functions in $(R^n, \wp_{R^n})$, the likelihood function of $\boldsymbol{\theta}$ for a given sample $\mathbf{x}$ is:

$$L(\boldsymbol{\theta}; \mathbf{x}) = kp_{\mathbf{x}}(\mathbf{x}; \boldsymbol{\theta}) \tag{5.7}$$

Here $k > 0$ is independent with the value of $\boldsymbol{\theta}$, usually $k = 1$. If statistics $\hat{\boldsymbol{\theta}}(x)$ exists from $(R^n, \wp_{R^n})$ to $(\Theta, \wp_{\Theta})$ and

$$L\left(\hat{\boldsymbol{\theta}}(\mathbf{x}); \mathbf{x}\right) = \sup_{\boldsymbol{\theta}} L(\boldsymbol{\theta}; \mathbf{x}) \tag{5.8}$$

$\hat{\boldsymbol{\theta}}(x)$ denotes the Maximum Likelihood Estimate (MLE) of $\boldsymbol{\theta}$. The logarithm of likelihood function can simplify the calculation process because probability intensity functions are often shown by exponential forms.

$$l(\boldsymbol{\theta}; \mathbf{x}) = \ln L(\boldsymbol{\theta}; \mathbf{x}) \tag{5.9}$$

$l(\boldsymbol{\theta}; \mathbf{x})$ is called a log-likelihood function. As logarithmic transformation is strictly monotone increasing, so $l(\boldsymbol{\theta}; \mathbf{x})$ and $\mathbf{L}(\boldsymbol{\theta}; \mathbf{x})$ have the same maximum. The most common used method to acquire MLE is derivation. If $\hat{\boldsymbol{\theta}}(x)$ is the interior point of Θ, $\hat{\boldsymbol{\theta}}(x)$ will be the solver of the following likelihood equation.

$$\partial l(\boldsymbol{\theta}; \mathbf{x})/\partial \theta_i = 0, \quad i = 1 \ldots k \tag{5.10}$$

5.2.4 Likelihood Ratio Test

(1) Binary hypothesis test

There are only original hypothesis H_0 and alternative hypothesis H_1 in binary hypothesis test. Many complex multiple hypothesis test can be decomposed as several basic binary hypothesis tests [4]. The observation model of binary hypothesis test is shown in Fig. 5.2. Two types of samples respectively come from original hypothesis H_0 and alternative hypothesis H_1. Observation space $\mathbf{Z}$ is divided into two disjoint observational sets according to testing criterion:

$$\mathbf{Z} = \mathbf{Z}_0 \cup \mathbf{Z}_1 \tag{5.11}$$

Observation value is judged coming from original hypothesis H_0 when it is in $\mathbf{Z}_0$ which called the rejection region, and observation value is judged coming from alternative hypothesis H_1 when it is in $\mathbf{Z}_1$.

The following situations will be referred in binary hypothesis test problem:

(1) Judge H_0, H_0 is true.
(2) Judge H_0, H_1 is true.

Fig. 5.2 Observation model
of binary hypothesis test [4]

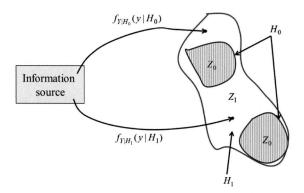

(3) Judge H_1, H_0 is true.
(4) Judge H_1, H_1 is true.

Situations (1) and (4) are correct judgments and situations (2) and (3) denote false judgements. (2) is called undetected, and (3) is called false alarm.

(2) Bayesian criterion

Bayesian criterion is utilized when prior probability $P(H_0)$, $P(H_1)$ and decision cost C_{00}, C_{10}, C_{11}, C_{01} are known. Generally, the cost of false judgments is always higher than true ones, namely $C_{00} < C_{10}$, $C_{11} < C_{01}$. The sum of prior probability is:

$$P(H_0) + P(H_1) = 1 \tag{5.12}$$

With given joint probability $P(D_i, H_j)(i, j = 0, 1)$, the average cost is:

$$R = E(C) = C_{00}P(D_0, H_0) + C_{00}P(D_0, H_1) + C_{00}P(D_1, H_0) + C_{00}P(D_1, H_1) \tag{5.13}$$

Using the Bayes formula

$$P(D_i, H_j) = P(D_i|H_j)P(H_j) \tag{5.14}$$

The probability of correct judgment is

$$\begin{aligned} P(c) &= P(D_0, H_0) + P(D_1, H_1) \\ &= (1 - P_F)P(H_0) + P_D P(H_1) \end{aligned} \tag{5.15}$$

And the probability of false judgment is

$$\begin{aligned} P(e) &= P(D_0, H_1) + P(D_1, H_0) \\ &= P_M P(H_1) + P_F P(H_0) \end{aligned} \tag{5.16}$$

So the average cost is

$$
\begin{aligned}
R &= E(C) \\
&= C_{00}(1 - P_F)P(H_0) + C_{01}(1 - P_D)P(H_1) + C_{10}P_F P(H_0) + C_{11}P_D P(H_1) \\
&= C_{00}P(H_0) \int_{Z_0} f_{\mathbf{x}|H_0}(\mathbf{x}|H_0)d\mathbf{x} + C_{01}P(H_1) \int_{Z_0} f_{\mathbf{x}|H_1}(\mathbf{x}|H_1)d\mathbf{x} \\
&\quad + C_{10}P(H_0) \int_{Z_1} f_{\mathbf{x}|H_0}(\mathbf{x}|H_0)d\mathbf{x} + C_{11}P(H_1) \int_{Z_1} f_{\mathbf{x}|H_1}(\mathbf{x}|H_1)d\mathbf{x}
\end{aligned}
$$

$$(5.17)$$

As

$$
\int_Z f_{\mathbf{x}|H_0}(\mathbf{x}|H_0)d\mathbf{x} = \int_z f_{\mathbf{x}|H_1}(\mathbf{x}|H_1)d\mathbf{x} = 1 \tag{5.18}
$$

and

$$
\int_{z_1} f_{\mathbf{x}|H_j}(\mathbf{x}|H_j)d\mathbf{x} = 1 - \int_{z_0} f_{\mathbf{x}|H_j}(\mathbf{x}|H_j)d\mathbf{x} \quad j = 0, 1 \tag{5.19}
$$

where $f_{\mathbf{x}|H_j}(\mathbf{x}|H_j)$ $(j = 0, 1)$ is the conditional probability of each hypothesis, (5.20) can be obtained by the substitute (5.19) into (5.17).

$$
\begin{aligned}
R &= P_0 C_{10} + P_1 C_{11} \\
&= \int_{z_0} ((P_1(C_{01} - C_{11})f_{\mathbf{x}|H_1}(\mathbf{x}|H_1)) - (P_0(C_{10} - C_{00})f_{\mathbf{x}|H_0}(\mathbf{x}|H_0)))d\mathbf{x}
\end{aligned}
$$

$$(5.20)$$

Constant $P_0 C_{10} + P_1 C_{11}$ is independent of the samples in the measurement space. As $(P_1(C_{01} - C_{11})f_{\mathbf{x}|H_1}(\mathbf{x}|H_1))$ and $P_0(C_{10} - C_{00})f_{\mathbf{x}|H_0}(\mathbf{x}|H_0)$ are both positive, the latter should be larger than the former in integration domain to make negative integration value and the lowest risk. The samples in Z_0 should satisfy:

$$
P_1(C_{01} - C_{11})f_{\mathbf{x}|H_1}(\mathbf{x}|H_1) < P_0(C_0 - C_{00})f_{\mathbf{x}|H_0}(\mathbf{x}|H_0) \tag{5.21}
$$

Samples which can make the second term larger should belong to Z_1. Samples which can make the two terms equivalent can be assigned to either Z_0 or Z_1 If

$$
(P_1(C_{01} - C_{11})f_{\mathbf{x}|H_1}(\mathbf{x}|H_1) > P_0(C_{10} - C_{00})f_{\mathbf{x}|H_0}(\mathbf{x}|H_0) \tag{5.22}
$$

Hypothesis H_1 is true. Otherwise, hypothesis H_0 is true. Thus, Bayes rules can be represented as:

$$\frac{f_{\mathbf{x}|H_1}(\mathbf{x}|H_1)}{f_{\mathbf{x}|H_0}(\mathbf{x}|H_0)} \begin{cases} > \frac{P_0(C_{10}-C_{00})}{P_1(C_{01}-C_{11})} & H_1 \text{ is true} \\ < \frac{P_0(C_{10}-C_{00})}{P_1(C_{01}-C_{11})} & H_0 \text{ is true} \end{cases} \tag{5.23}$$

Risk function is simplified when error cost is 1 and correct cost is 0

$$R = E(C) = P_M P_1 + P_F P_0 = P(e) \tag{5.24}$$

Decision threshold is simplified when minimum average cost equals with minimum error probability.

$$\eta = \frac{P_0}{P_1} \tag{5.25}$$

(3) Neyman-Pearson Criterion

Neyman-Pearson criterion can be used without cost and prior probability, which makes detection probability P_D maximal when false alarm probability P_F is fixed. Maximal P_D is equivalent to minimum P_M because $P_M = 1 - P_D$ [4]. In order to make P_M minimum or P_D maximum with $P_F = \alpha$, objective function is established

$$J = P_M + \lambda(P_F - \alpha) \tag{5.26}$$

where $\lambda(\lambda \geq 0)$ is called the Lagrange multiplier. There are several decision space Z_1 which satisfy with $P_F = \alpha$, so The objective function should be used to select a proper Z_1 to make P_M minimum.

$$\begin{aligned} J &= \int_{Z_0} f_{\mathbf{x}|H_1}(\mathbf{x}|H_1)d\mathbf{x} + \lambda[\int_{Z_1} f_{\mathbf{x}|H_0}(\mathbf{x}|H_0)d\mathbf{x} - \alpha] \\ &= 1 - \int_{Z_1} f_{\mathbf{x}|H_1}(\mathbf{x}|H_1)d\mathbf{x} + \lambda \int_{Z_1} f_{\mathbf{x}|H_0}(\mathbf{x}|H_0)d\mathbf{x} - \lambda\alpha \\ &= (1 - \lambda\alpha) + \int_{Z_1} (\lambda f_{\mathbf{x}|H_0}(\mathbf{x}|H_0) - f_{\mathbf{x}|H_1}(\mathbf{x}|H_1))d\mathbf{x} \end{aligned} \tag{5.27}$$

Hence, the value of J is the minimum when integration term $\int_{Z_1}[\lambda f_{\mathbf{x}|H_0}(\mathbf{x}|H_0) - f_{\mathbf{x}|H_1}(\mathbf{x}|H_1)]d\mathbf{x}$ is minimum, the decision criterion is represented:

$$\Lambda(\mathbf{x}) = \frac{f_{\mathbf{x}|H_1}(\mathbf{x}|H_1)}{f_{\mathbf{x}|H_0}(\mathbf{x}|H_0)} \begin{cases} > \lambda & H_1 \text{ is true} \\ < \lambda & H_0 \text{ is true} \end{cases} \tag{5.28}$$

(4) Generalized likelihood ratio test

Likelihood ratio cannot directly acquire from measurements when unknown parameters are contained in likelihood function. So generalized likelihood ratio is applied by substituting the maximum likelihood estimation of unknown parameters into the formula of likelihood ratio [4].

$$\Lambda_G(\mathbf{x}) = \frac{\sup\limits_{\boldsymbol{\theta} \in \Theta_1} L(\mathbf{x}; \boldsymbol{\theta})}{\sup\limits_{\boldsymbol{\theta} \in \Theta_0} L(\mathbf{x}; \boldsymbol{\theta})} = \frac{L(\mathbf{x}; \hat{\boldsymbol{\theta}}_1)}{L(\mathbf{x}; \hat{\boldsymbol{\theta}}_0)} \tag{5.29}$$

Then the corresponding likelihood ratio is

$$\Lambda_G(\mathbf{x}) = \begin{cases} > \lambda & H_1 \text{ is true} \\ < \lambda & H_0 \text{ is true} \end{cases} \tag{5.30}$$

where λ is a threshold.

5.2.5 Subspace Analysis

(1) Projection operator and projection matrix

Assuming that $\mathbf{L}$ and $\mathbf{M}$ are subspaces of $\mathbf{C}^n$, and $\mathbf{L} \oplus \mathbf{M} = \mathbf{C}^n$. An arbitrary $\mathbf{x} \in \mathbf{C}^n$ can be uniquely decomposed.

$$\mathbf{x} = \mathbf{y} + \mathbf{z}, \quad \mathbf{y} \in \mathbf{L}, \quad \mathbf{z} \in \mathbf{M} \tag{5.31}$$

where $\mathbf{y}$ is the projection of $\mathbf{x}$ along $\mathbf{M}$ to $\mathbf{L}$.

Definition 5.8 Projection operator $\mathbf{P}_{LM}$ along $\mathbf{M}$ to $\mathbf{L}$ is defined as the transformation of an arbitrary $\mathbf{x} \in \mathbf{C}^n$ projection along subspace $\mathbf{M}$ to $\mathbf{L}$.

$$\mathbf{P}_{LM}\mathbf{x} = \mathbf{y} \tag{5.32}$$

The matrix of the vector base $\mathbf{e}_1, \mathbf{e}_2, \ldots, \mathbf{e}_n$ of $\mathbf{P}_{LM}$ in $\mathbf{C}^n$ is called the projection matrix, which is denoted as $\mathbf{P}_{LM}$ [6]. Projection matrix $\mathbf{P}_{LM}$ can be obtained by the following method. Assuming $\dim \mathbf{L} = r$, $\dim \mathbf{M} = n - r$. The vector bases in subspace $\mathbf{M}$ and $\mathbf{L}$ $\{\mathbf{x}_1, \mathbf{x}_2 \ldots \mathbf{x}_{n-r}\}$, $\{\mathbf{y}_1, \mathbf{y}_2 \ldots \mathbf{y}_{n-r}\}$ are jointly composed as the vector base of $\mathbf{C}^n$. According to the characteristic of projection matrix

$$\begin{aligned} \mathbf{P}_{L,M}\mathbf{x}_i &= \mathbf{x}_i \quad (i = 1, 2, \ldots r) \\ \mathbf{P}_{L,M}\mathbf{y}_j &= \mathbf{0} \quad (j = 1, 2, \ldots, n - r) \end{aligned} \tag{5.33}$$

There are $\mathbf{X} = \{\mathbf{x}_1, \mathbf{x}_2 \ldots \mathbf{x}_{n-r}\}$ and $\mathbf{Y} = \{\mathbf{y}_1, \mathbf{y}_2 \ldots \mathbf{y}_{n-r}\}$, (5.31) is equivalent to

$$P_{L,M}[\mathbf{X}\!:\!\mathbf{Y}] = [\mathbf{X}\!:\!\mathbf{O}] \tag{5.34}$$

As $[\mathbf{X}\!:\!\mathbf{Y}]$ is an n order invertible matrix, projection matrix is

$$P_{L,M}[\mathbf{X}\!:\!\mathbf{O}] = [\mathbf{X}\!:\!\mathbf{Y}]^{-1} \tag{5.35}$$

(2) Orthogonal projection operator and orthogonal projection matrix.

Definition 5.9 If $\mathbf{x}$ and all of vectors in subspace $\mathbf{L}$ are orthogonal, $\mathbf{x}$ and $\mathbf{L}$ are also orthogonal; if arbitrary vectors in subspace $\mathbf{L}_1$ and $\mathbf{L}_2$ are orthogonal, $\mathbf{L}_1$ and $\mathbf{L}_2$ are also orthogonal. $\forall \mathbf{a}_i \in \mathbf{L}_i, \forall \mathbf{a}_j \in \mathbf{L}_j, \mathbf{a}_i \perp \mathbf{a}_j \Rightarrow \mathbf{L}_i \perp \mathbf{L}_j$.

Definition 5.10 The subspace $\mathbf{L}^\perp$ composed of all of vectors which are orthogonal with $\mathbf{L}$ is called the orthogonal complement subspace of $\mathbf{L}$, $\mathbf{L}^\perp = \{\mathbf{x} | \mathbf{x}^T \mathbf{y} = 0, \forall \mathbf{y} \in \mathbf{L}\}$.

Definition 5.11 Assuming that $\mathbf{L}$ is a subspace of $\mathbf{C}^n$, the projection operator $P_{L,L^\perp}$ along $\mathbf{L}^\perp$ to $\mathbf{L}$ is called orthogonal projection operator which is simply denoted as P_L. And the matrix of the vector base $\mathbf{e}_1, \mathbf{e}_2, \ldots, \mathbf{e}_n$ of P_L in $\mathbf{C}^n$ is called the orthogonal projection matrix, which is denoted as P_L. Orthogonal projection matrix P_L can be obtain by the following method with known vector base of $\mathbf{L}$. Assuming that $\dim \mathbf{L} = r$, $\dim \mathbf{M} = n - r$. $\{\mathbf{x}_1, \mathbf{x}_2 \ldots \mathbf{x}_{n-r}\}$ is used as the vector base of $\mathbf{L}$ while $\{\mathbf{y}_1, \mathbf{y}_2 \cdots \mathbf{y}_{n-r}\}$ is assumed to be the vector base of $\mathbf{L}^\perp$. There are partitioned matrix $\mathbf{X} = \{\mathbf{x}_1, \mathbf{x}_2 \ldots \mathbf{x}_{n-r}\}$ and $\mathbf{Y} = \{\mathbf{y}_1, \mathbf{y}_2 \ldots \mathbf{y}_{n-r}\}$, obviously, $\mathbf{X}^H \mathbf{Y} = \mathbf{O}$. From (5.33), there is

$$\begin{aligned}
P_L &= [\mathbf{X}\!:\!\mathbf{O}][\mathbf{X}\!:\!\mathbf{Y}]^{-1} \\
&= [\mathbf{X}\!:\!\mathbf{O}][[\mathbf{X}\!:\!\mathbf{Y}]^{-H}[\mathbf{X}\!:\!\mathbf{Y}]]^{-1}[\mathbf{X}\!:\!\mathbf{Y}]^H \\
&= \mathbf{X}(\mathbf{X}^H \mathbf{X})^{-1} \mathbf{X}^H
\end{aligned} \tag{5.36}$$

Thus, if the measurement vector is used as vector base of $\mathbf{L}$, according to least square optimization criteria, the projection is shown as follows after operation of (5.34).

$$\hat{\mathbf{x}} = P_L \cdot \mathbf{x} \tag{5.37}$$

It is the least squares estimation of $\mathbf{x}$ [6].

5.2.6 Stochastic Signal and Entropy

Information is the message which may have effect on events conducted by the source. And source space is also called uncertain space. The degree of uncertainty can be described by a probability function of source space. The amount of information is related to the probability. The lower probability is, the larger amount of information. Shannon had used reciprocal of probability to describe the information of one event [7]. Assuming that the probability of event x is $p(x)$, the self-information of x is:

$$\Delta I(X = x) = \log\left(\frac{1}{p(x)}\right) = -\log(p(x)) \tag{5.38}$$

Entropy is defined as the expectation of information amount, which is used to describe overall characteristics of information source:

$$H = E(\Delta I) \tag{5.39}$$

Here, continuous and discrete random variables are discussed as follows:

(1) When x is a continuous random variable whose probability density function is $f(x)$, entropy is defined as:

$$H_f = -\int f(x)\log(f(x)) \tag{5.40}$$

(2) When x is a discrete random variable whose probability density function is $P(x)$, entropy is defined as:

$$H_P = -\sum_x P(x)\log(P(x)) \tag{5.41}$$

The following theorem can be proved when x is a continuous random variable [7].

Theorem 5.1 Maximum entropy theorem

Let $\mathbf{x} = (x_1, x_2, \ldots, x_p)^T$ has a multivariate normal distribution with given covariance matrix Σ, then $\max H_f = \frac{1}{2}\log(2\pi e)^n \det|\Sigma|$, where $|\Sigma|$ denotes the determinant of Σ.

Proof $H_f = -\int_{\mathbf{R}^n} f(\mathbf{x})\log f(\mathbf{x})d\mathbf{x} = -\int_{\mathbf{R}^n} f(\mathbf{x})\log Q(\mathbf{x})d\mathbf{x} + \int_{\mathbf{R}^n} f(\mathbf{x})\log\frac{Q(\mathbf{x})}{f(\mathbf{x})}d\mathbf{x}$ where $Q(\mathbf{x})$ is Gaussian intensity, $Q(\mathbf{x}) = \frac{1}{(2\pi)^{\frac{n}{2}}\det|\Sigma|}\exp\left\{-\frac{1}{2}(\mathbf{x}-\mathbf{m})\Sigma^{-1}(\mathbf{x}-\mathbf{m})^T\right\}$. Σ is covariance matrix, $n = 1, 2, \ldots$ is the dimension of observation vector, Thus

$$H_f \leq \int_{\mathbf{R}^n} f(x) \left[\frac{1}{2} \log(2\pi)^2 \det|\mathbf{\Sigma}| + \frac{1}{2}(\mathbf{x} - \mathbf{m})\mathbf{\Sigma}^{-1}(\mathbf{x} - \mathbf{m})^T \right] d\mathbf{x}$$

$$+ \int_{\mathbf{R}^n} f(x) \left[\frac{Q(\mathbf{x})}{f(\mathbf{x})} - 1 \right] d\mathbf{x} = \frac{1}{2} \log(2\pi e)^n \det|\mathbf{\Sigma}|$$

where the fundamental inequality $\log x \leq x - 1$ in information theory is utilized.
   The following theorem can be proved when x is a discrete random variable.

Theorem 5.2 H_p *reaches the maximum value* $\log N$ *when the probability all values of x are equivalent, where N denotes the number of values of x.*

Proof Utilize the fundamental inequality $\log x \leq x - 1$

$$H_P \left(P_X(x) = \frac{1}{N} \right) - H_P = -\sum_x \frac{1}{N} \log \frac{1}{N} + \sum_x P(x) \log(P(x))$$

$$= -\sum_x P(x) \log \frac{1}{NP(x)} \tag{5.42}$$

$$\geq -\sum_x P(x) \left(1 - \log \frac{1}{NP(x)} \right) = 0$$

5.2.7 Nonparametric Kernel Density Estimation

Random variable can be described through its probability density function. However, the probability density function is always unknown in practical analysis. The estimation of probability density function is an important problem to be solved. Here introduce three kinds of common estimation approaches [1, 8–10]:

(1) Parametric estimation: assuming that samples come from the probability density function of given models, which can be converted into a model parameter estimation problem based on samples data. However, due to the known form of probability density function in prior, model error will be induced because of the differences between practical and estimated probability density function.

(2) Nonparametric estimation (e.g. Parzen window estimator, also called Kernel Density Estimation). It is a numerical method without the prior of probability density function forms. Sample data provide all of information. When Gaussian kernel function is used, the probability density function is acquired using the average of Gaussian density of each sample.

(3) Semi-parametric estimation: it uses semi-parametric model without the prior of probability density function forms. However, semi-parametric model constraints the number of parameters, which means the complexity will be limited.

There are other probability density estimation methods which respectively based on spline, orthogonal basis and histogram. Nonparametric estimation method is mainly introduce [1, 8–10]. Assuming that there is a group of samples $x_1, \ldots, x_n$ from the unknown probability density f, the density estimation $\hat{f}_n$ is a Borel observational function of variable x and sample data $x_1, \ldots, x_n$:

$$\hat{f}_n = f_n(x, x_1, \ldots, x_n) \tag{5.43}$$

where $\hat{f}_n \geq 0$ and $\int_{\mathbf{R}} \hat{f}_n(x)dx = 1$. The probability measurement of f is denoted as F. λ is the Lebesgue measurement. So f is almost everywhere equal with the Radon-Nikodym differential $\frac{dF}{d\lambda}$ of F. The following formula can be acquired for all of Borel sets A:

$$\int_A f = F(A) \tag{5.44}$$

Actually many estimation methods aim to seek the optimal approximation of differential $\frac{dF}{d\lambda}$.

Moving window estimation was proposed by Rosenblatt [1, 8].

$$\hat{f}_n(x) = \frac{F_n(S(x, h))}{\lambda(S(x, h))} \tag{5.45}$$

where $S(x, h)$ is a sphere whose center is x and diameter is h. h is also called smoothing factor. In real domain, $\hat{f}_n(x)$ calculates the number of samples in $[x - h, x + h]$ and then divides $2hn$. When n is large enough and h is enough small, $\hat{f}_n(x)$ is approximated to

$$\hat{f}_n(x) \approx \frac{F(S(x, h))}{\lambda(S(x, h))} \tag{5.46}$$

$\hat{f}_n(x)$ is approximated to real distribution f. Window function $S(x, h)$ is actually equivalent to a density function with uniform distribution. So a more general function (kernel function) K is used to replace S to acquire the basic kernel density estimator:

$$f_n(x) = \frac{1}{n} \sum_{i=1}^{n} K_h(x - x_i) \tag{5.47}$$

Table 5.1 Commonly used 2nd-order kernel functions

Kernel functions	$K(u), \delta(|u| \leq 1)$
Uniform	$\frac{1}{2}\delta(|u| \leq 1)$
Triangle	$(1-|u|)\delta(|u| \leq 1)$
Epanechnikov	$\frac{3}{4}(1-u^2)\delta(|u| \leq 1)$
Quartic	$\frac{15}{16}(1-u^2)^2\delta(|u| \leq 1)$
Triweight	$\frac{35}{32}(1-u^2)^2\delta(|u| \leq 1)$
Gaussian	$\frac{1}{\sqrt{2\pi}}\exp(-\frac{1}{2}u^2)$

where $K_h(x) = \frac{1}{h}K(\frac{x}{h})$. K is symmetric, $K(u) = K(-u)$. And a k order kernel function satisfies:

$$\int_{\mathbf{R}} K(u)du = 1$$

$$\int_{\mathbf{R}} u^j K(u)du = 0, \quad j = 1, \ldots, k-1 \tag{5.48}$$

$$\int_{\mathbf{R}} u^k K(u)du \neq 0$$

Table 5.1 shows several common used two order kernel functions [10]. It can be extended into p-dimension.

$$\hat{f}_n(\mathbf{x}) = \hat{f}_n(x_1, x_2, \ldots, x_p)$$

$$= \frac{1}{nh_1\ldots h_p}\sum_{i=1}^{n} K_{h_1,\ldots,h_p}\left[\frac{x_1 - X_{1i}}{h_1}, \frac{x_2 - X_{2i}}{h_2}, \ldots, \frac{x_p - X_{pi}}{h_p}\right] \tag{5.49}$$

where $\mathbf{x} = [x_1, x_2, \ldots, x_p]$, and $K_{h_1}, \ldots, _{h_p}$ is a vector kernel function.

$$K_{h_1}, \ldots, _{h_p}(y) = \prod_{i=1}^{p} K_{h_i}(y_i) \tag{5.50}$$

$K_{h_i}(\cdot)$ is a scalar kernel function.

5.3 Anomaly Detection Methods

Anomaly detection methods are used to detect the targets in polarized spectrum images when there are not prior information about the spectral characteristics of targets. This method is independent of multi-band polarization prior information, which can judge the pixels inconsistent with the background features as the target.

However, it is difficult to apply the multi-band polarization prior information in practical. Firstly, the application environment can be so complex that the cost of acquiring prior multi-band polarization features is very high. Secondly, the measured multi-band polarization features are always uncertain in practical applications, which need stable inverse model or algorithms or other assistant processing. Finally, it is a large cost to do plenty of mapping experiments before setting a practical multi-band polarization features library. Thus, multi-band polarization prior information is proofed to be unsuitable in practical applications.

There are two types of commonly used high dimension anomaly detection methods, one judges the anomaly points inconsistent with background statistical characteristics as targets, which is represented by Reed's RXD algorithm [11–13] and UTD algorithm [3, 14]. RXD is used to detect anomaly points based on the spatial whitening normal stochastic process. Another one is used to seek background features in high dimension measurement data and then detect anomaly points using subspace projection or other matching methods [6] which is represented by the LPD algorithm that the principle components are used as background features to achieve orthogonal subspace projection.

5.3.1 Statistical Hypothesis Model for Anomaly Detection

Multi-band polarization mixture model is not generally referred without corresponding prior information. As for the high dimension measurement data with B bands and N pixels, a B dimension vector $\mathbf{x} = [x_1, x_2 \ldots x_B]$ is used to represent the arbitrary pixel position, and a hypothesis model for high dimension data is as follows:

$$\begin{cases} H_0 : \mathbf{x} \sim p_0(\mathbf{x}) \\ H_1 : \mathbf{x} \sim p_1(\mathbf{x}) \end{cases} \tag{5.51}$$

where $\mathbf{x}$ is the pixel to be detect, $p_0(\mathbf{x})$ and $p_1(\mathbf{x})$ are different statistical distribution which respectively represented the statistical characteristics of background signal s and target signals. The model in (5.51) can be represented the following additional forms:

$$\begin{cases} H_0 : \mathbf{x} = \mathbf{x}_b + \mathbf{v} \\ H_1 : \mathbf{x} = \mathbf{x}_t + \mathbf{x}_b + \mathbf{v} \end{cases} \tag{5.52}$$

and

$$\begin{cases} H_0 : \mathbf{x} = \mathbf{c} \\ H_1 : \mathbf{x} = \mathbf{x}_t + \mathbf{c} \end{cases} \tag{5.53}$$

where $\mathbf{x}_t$ is target signal, $\mathbf{x}_b$ is background signal. $\mathbf{v}$ is noise signal. $\mathbf{c}$ is background clutter. Structured and unstructured background signals are respectively used in (5.52) and (5.53) [3].

5.3.2   Low Probability Target Detection

Low probability target detection (LPD) is based on orthogonal subspace projection. It acquires background information in an unsupervised way and then inhibits background information using orthogonal projection, and reduces the information of low probability targets to enlarge the differences between targets and background. Assuming that matrix $\mathbf{V} = \left[\mathbf{v}_1, \mathbf{v}_2 \ldots, \mathbf{v}_i, \ldots, \mathbf{v}_q\right]$ represents the first q principal components of high dimension data which embody background information. Orthogonal projection operator $\mathbf{P}$ is constructed by $\mathbf{V}$.

$$\mathbf{P} = \left(\mathbf{I}_{B \times B} - \mathbf{V}\mathbf{V}^{\#}\right) \tag{5.54}$$

$\mathbf{V}^{\#} = \left(\mathbf{V}^T \mathbf{V}\right)^{-1} \mathbf{V}^T$ is the pseudo-inverse of $\mathbf{V}$. $\mathbf{P}$ is a $B \times B$ matrix. $\mathbf{I}_{B \times B}$ is a B order unit matrix. $\mathbf{M}$ Projection of pixel $\mathbf{x}$ is acquired by projection operator $\mathbf{P}$ to get q bands images after orthogonal projection, where the data of an arbitrary pixel is represented by a B dimension vector $\tilde{\mathbf{x}}$.

$$\tilde{\mathbf{x}} = \mathbf{P} \cdot \mathbf{x} \tag{5.55}$$

Output PSNR will be the highest when projection is along the direction of target signal [6]. Here is the detection data $\mathbf{R}$:

$$\mathbf{R} = \hat{\mathbf{d}}^T \tilde{\mathbf{x}}_i = \hat{\mathbf{d}}^T \mathbf{P} \cdot \mathbf{x} \tag{5.56}$$

where $\hat{\mathbf{d}}^T$ is estimation values of target, $\hat{\mathbf{d}} = \left[11 \ldots 1\right]^T$. The choice of target estimation is not capable of inhibiting the remained background information which will lead to false alarm, undetected and other problems.

5.3.3   Classical RX Detection Algorithm

RX algorithm is derived and progressive simplified through generalized likelihood ratio test on the basic of whitening Gaussian distribution [11–13]:

$$\delta(\mathbf{x}) = (\mathbf{x} - \boldsymbol{\mu})^T (\frac{N}{N+1}\hat{\boldsymbol{\Sigma}} + \frac{1}{N+1}(\mathbf{x} - \boldsymbol{\mu})(\mathbf{x} - \boldsymbol{\mu})^T)^{-1}(\mathbf{x} - \boldsymbol{\mu})$$

$$\begin{cases} \geq \lambda & \text{Target} \\ < \lambda & \text{Background} \end{cases} \tag{5.57}$$

(5.57) can be simplified:

$$\delta_{RXD}(\mathbf{x}) = (\mathbf{x} - \boldsymbol{\mu})^T \mathbf{M}^{-1}(\mathbf{x} - \boldsymbol{\mu}) \begin{cases} \geq \lambda & \text{Target} \\ \leq \lambda & \text{Background} \end{cases} \tag{5.58}$$

where $\boldsymbol{\mu} = \frac{1}{N}\sum_{n=1}^{N}\mathbf{x}_i$ is sample average of multi-band polarization image data, $\mathbf{M} = \frac{1}{N}\sum_{n=1}^{N}(\mathbf{x}_i - \boldsymbol{\mu})(\mathbf{x}_i - \boldsymbol{\mu})^T$ is the covariance matrix.

5.3.4   Improved RX Detection Algorithm

An expression of RX algorithm is similar to the matching filter:

$$M_d(\mathbf{x}) = k \cdot \mathbf{d}^T \mathbf{x} \tag{5.59}$$

where $\mathbf{d}$ is matching signal, k is a constant or the function of $\mathbf{x}$. The performance of detector is determined by $\mathbf{d}$ and $\mathbf{x}$. Thus, the standard RX algorithm can be considered as a detector for $\mathbf{x} - \boldsymbol{\mu}$ where $k = 1$ and $\mathbf{d}^T = (\mathbf{x} - \boldsymbol{\mu})^T \mathbf{M}^{-1}$. Different matching $\mathbf{d}$ and constant k are used to improved RX algorithm:

$$\delta_{MRXD}(\mathbf{x}) = \bar{\mathbf{x}}^T \mathbf{M}^{-1}(\mathbf{x} - \boldsymbol{\mu}) \tag{5.60}$$

$$\delta_{NRXD}(\mathbf{x}) = \bar{\mathbf{x}}^T \mathbf{M}^{-1}\bar{\mathbf{x}} \tag{5.61}$$

where $\bar{\mathbf{x}} = (\mathbf{x} - \boldsymbol{\mu})/\|\mathbf{x} - \boldsymbol{\mu}\|$, $\|\mathbf{x} - \boldsymbol{\mu}\| = ((\mathbf{x} - \boldsymbol{\mu})^T(\mathbf{x} - \boldsymbol{\mu}))^{1/2}$. The one order statistical information will be left out by the covariance matrix $\mathbf{M}$ in (5.60) and (5.61), which may probably affect detection performance. Replacing $\mathbf{M}$ and $\mathbf{x} - \boldsymbol{\mu}$ respectively by dependency matrix $\mathbf{R}$ and $\mathbf{x}$, three kinds of improved RX algorithm can be acquired:

$$\delta_{CRRXD}(\mathbf{x}) = \mathbf{x}^T \mathbf{R}^{-1}\mathbf{x} \tag{5.62}$$

$$\delta_{CRMRXD}(\mathbf{x}) = \left(\frac{\mathbf{x}}{\|\mathbf{x}\|}\right)^T \mathbf{R}^{-1}\mathbf{x} \tag{5.63}$$

$$\delta_{CRNRXD}(\mathbf{x}) = \left(\frac{\mathbf{x}}{\|\mathbf{x}\|}\right)^T \mathbf{R}^{-1}\left(\frac{\mathbf{x}}{\|\mathbf{x}\|}\right) \tag{5.64}$$

5.4   Fuzzy Integral Fusion Detection of Band Subsets

5.4.1   Data Model and Band Subset Division

Let N denote the amount of pixel samples of multi-band polarization data cubes, and L denotes the amount of spectral bands. The multi-band polarization data cube can be represented by a $N \times L$ matrix $\mathbf{X} = [\mathbf{x}(1), \mathbf{x}(2), \ldots, \mathbf{x}(k), \ldots, \mathbf{x}(N)]^T$. $\mathbf{x}(k)$ is the kth pixel vector where every element represents multi-band polarization response in different bands. From the view of object detection, $\mathbf{X}$ can be considered as a linear mixture of target signals, background signals and the noise:

$$\mathbf{X} = \mathbf{S} + \mathbf{W} = \mathbf{B} + \mathbf{T} + \mathbf{W} \tag{5.65}$$

$\mathbf{S} = \mathbf{B} + \mathbf{T} = [\mathbf{s}(1), \mathbf{s}(2), \ldots, \mathbf{s}(k), \ldots, \mathbf{s}(N)]^T$ denotes signal matrix which contains the background signal $\mathbf{B} = [\mathbf{b}(1), \mathbf{b}(2), \ldots, \mathbf{b}(k), \ldots, \mathbf{b}(N)]^T$ and the target signal $\mathbf{T} = [\mathbf{t}(1), \mathbf{t}(2), \ldots, \mathbf{t}(k), \ldots, \mathbf{t}(N)]^T$. $\mathbf{s}(k)$, $\mathbf{b}(k)$ and $\mathbf{t}(k)$ respectively represent data signal, background signal and target signal at the kth pixel position. If there is no target information contained in pixel k, $\mathbf{t}(k) = 0$. $\mathbf{T}$ is actually a sparse matrix for that target information occupies only few pixels in practical anomaly detection. $\mathbf{W} = [\boldsymbol{\omega}_1, \boldsymbol{\omega}_2, \ldots, \boldsymbol{\omega}_l, \ldots, \boldsymbol{\omega}_L]$ is the noise matrix where $\boldsymbol{\omega}_1$ is the $N \times 1$ noise vector of the lth bands.

A simplified form of RX algorithm:

$$\delta(\mathbf{x}) = (\mathbf{x} - \mathbf{u})^T \hat{\boldsymbol{\Sigma}}^{-1} (\mathbf{x} - \mathbf{u}) \overset{H_1}{\underset{H_0}{\gtrless}} \eta \tag{5.66}$$

$\hat{\boldsymbol{\Sigma}} = \frac{1}{N} \sum_{k=1}^{N} (\mathbf{x}(k) - \mathbf{u})(\mathbf{x}(k) - \mathbf{u})^T$ is the covariance matrix, $\mathbf{u}$ is sample mean, η is the detection threshold which can be acquired through Neyman Pearson criterion.

The electromagnetic properties of different bands are always inconsistent, as well as the transmission characteristics and sensor noise [15–17]. These differences will be ignored when input the data into detector without any distinction, however, some of poor quality bands such as noise bands will have effects on the final results. Moreover, if all of bands are used to estimated covariance, there will be some problems, such as poor estimation accuracy and singularity problem, due to the high data dimension. To solve this problem, we divide the data into several subsets according to the bands similarity which is defined by the relative coefficient. The data in every subset has high similarity and low dimension. The relative coefficient between i-th and j-th band is defined as $\mathbf{R}_{corr}(i,j) = \mathbf{C}(i,j) / \sqrt{\mathbf{C}(i,i)\mathbf{C}(j,j)}$, where $\mathbf{C}$ is the $L \times L$ covariance matrix. All of bands can be divided into Q subsets according to the relativity. There will be continuous bands with high relativity in each subset. Let l_q denote the amount of bands in the qth subset. And apply the detection methods in Sect. 5.2 on each subset to acquire detection results which has complementarity and redundancy. More reliable results can be acquired by the fusion of all. Here, RXD algorithm is taken as an example to apply in the subset with Q bands, the detection results are denoted as $\delta_1(\cdot), \delta_2(\cdot), \ldots, \delta_q(\cdot), \ldots, \delta_Q(\cdot)$.

5.4.2 Fuzzy Measure and Fuzzy Integral

Fuzzy measurement was proposed in 1970s [18–21]. Here is the standard definition of fuzzy measure:

Let $\langle X, h, g \rangle$ denote a fuzzy measure space where $X = \{x_1, x_2, \ldots, x_n\}$ is a nonempty finite set. And $h : X \rightarrow [0, 1]$ denotes the support function on X ranging from 0 to 1. For given $x \in X$, $h(x)$ is the confidence. And $\Omega(X)$ is the power set of X, assuming that $h(x_1) \geq h(x_2) \geq \cdots h(x_n)$, $g : \Omega(X) \rightarrow [0, 1]$ will be the fuzzy measure with the following characteristics.

(1) Boundary condition: $g(\varphi) = 0, g(X) = 1$;
(2) Monotonicity: if $A, B \subset \Omega$ and $A \subset B$, $g(A) \leq g(B)$;
(3) Continuity: if $A_i \in \Omega, 1 \leq i \leq \infty$, and $\{A_i\}$ is monotonic, $\lim_{i \to \infty} g(A_i)$ $= g(\lim_{i \to \infty} A_i)$.

Fuzzy integral has successfully applied in many fields due to the fusion of information sources with different confidence [21]. In this section, fuzzy integral is used to fuse the detection results of all subsets. Let $\{X, h, g\}_y$ ($X = \{x_1(\mathbf{y}_1), x_2(\mathbf{y}_2), \ldots x_q(\mathbf{y}_q), \ldots x_Q(\mathbf{y}_Q)\}$) denote the fuzzy measure space of each given sample y. $\mathbf{y}_q$ denotes a subvector belonging to the q-th subset which has detection output $x_q(\mathbf{y}_q) = \delta_q(\mathbf{y}_q), (q = 1, 2, \ldots, Q)$. $h(x_q(\mathbf{y}_q))$ is the evident of target. Rearrange $x_q(\mathbf{y}_q)$ to satisfy $h(x_1(\mathbf{y}_1)) \geq h(x_2(\mathbf{y}_2)) \geq \cdots h(x_Q(\mathbf{y}_Q))$. And the Sugeno integral and Choquet integral about fuzzy measure are:

$$
\Lambda_s(\mathbf{y}) = \int_X h(x_q(\mathbf{y}_q)) \circ g(\cdot)
$$
$$
= \max_{\substack{q=1 \\ A_q \subseteq X}}^{Q} \left\{ \min \left[h(x_q(\mathbf{y}_q)), g(A_q) \right] \right\}
$$

(5.67)

And

$$
\Lambda_C(\mathbf{y}) = \int_X h(x_q(\mathbf{y}_q)) \circ g(\cdot)
$$
$$
= \max_{\substack{q=1 \\ A_q \subseteq X}}^{Q} \left[h(x_q(\mathbf{y}_q)) - h(x_{q+1}(\mathbf{y}_{q+1})) \right] g(A_q)
$$

(5.68)

$A_q = \{x_1(\mathbf{y}_1), x_2(\mathbf{y}_2), \ldots, x_q(\mathbf{y}_q)\}$ is an increasing subset sequence of X, $A \in \Omega(X)$ and $h(x_{Q+1}) = 0$. The operator $\circ$ means combination. If $g(\cdot)$ is the fuzzy measure of g_λ, $g(A_q)$ can be acquired by the following recursive process:

$$g(A_1) = g(\{x_1(\mathbf{y}_1)\}) = g_1 \tag{5.69}$$

$$g(A_q) = g_q + g(A_{q-1}) + \lambda g_q g(A_{q-1}) \tag{5.70}$$

$g_q = g_\lambda(\{x_q(\mathbf{y}_q)\})$ is fuzzy density which satisfies with $0 \leq g_q \leq 1$, $\lambda \in (-1, \infty)$ and $\lambda \neq 0$. Fuzzy density g_q can be considered as the importance of the q-th bands subset to final results. λ can be uniquely solved by [18, 21]:

$$\lambda + 1 \prod_{q=1}^{Q} (1 + \lambda g_q) \tag{5.71}$$

Every sample $\mathbf{y}(k)(k = 1, 2, \ldots, N)$ corresponds to a fuzzy measure space $\{X, h, g\}_{y(k)}$. Being similar with the RXD detector in (5.57), the Sugeno/Choquet fuzzy integral fusion detector are shown:

$$\Lambda_s(\mathbf{y}) = \max_{q=1}^{Q} \min[h(x_q(\mathbf{y}_q)), g(A_q)] \begin{cases} \geq \xi_s & H_1 : \text{Target present} \\ < \xi_s & H_0 : \text{Target absent} \end{cases} \tag{5.72}$$

$$\Lambda_c(\mathbf{y}) = \sum_{q=1}^{Q} [h(x_q(\mathbf{y}_q)) - h(x_{q+1}(\mathbf{y}_{q+1}))]g(A_q) \begin{cases} \geq \xi_c & H_1 : \text{Target present} \\ < \xi_c & H_0 : \text{Target absent} \end{cases} \tag{5.73}$$

where ξ_s and ξ_c are the corresponding detection thresholds.

5.4.3   Nonparametric Fuzzy Membership Functions

Before fuzzy integral fusion, support function $h(x_q(\mathbf{y}_q))$ $(q = 1, 2, \ldots, Q)$ should be estimated and fuzzy density g_q $(q = 1, 2, \ldots Q)$ of each subset should be set. As there are few samples, all of samples can be seen as coming from the background. RX detector calculates the Mahalanobis distance between detecting sample and all of samples, so the output of RX can be seen as the distance between detecting samples to the background. The probability of being target is higher when the value of output is higher. Thus, the output is considered as a measurement which is normalized to the range of [0,1]. However, simply normalized methods such as the linear stretch method only concerns the maximum and minimum values without using all of information. As probability density can contain all of information, we can detect the output of every subset to estimate their probability density functions.

But the modeled parametric estimation always has serious error due to the complexity of practical data. It is shown that multi-band polarization data have obvious long tail phenomenon and always be inconsistent with Gaussian model, so Chisquare distribution is not suitable for modeling output of RX detector [3]. In this section, a nonparametric kernel density estimation method [1, 8–10] is used to get the statistical characteristics of Q subsets detection. It is a numerical method without any prior hypothesis about probability distribution.

Let $\hat{f}_q$ be the probability density function of the q-th subset detection output $\delta_q(\cdot)$ by kernel density estimation. $\hat{f}_q(\delta_q(\mathbf{y}_q(k)))$ is the value of probability density of corresponding sample $\mathbf{y}_q(k)$. Target samples are sparse as they are quite fewer than background samples, so the effects on probability density function estimation are neglected. And $\hat{f}_q$ is approximately considered as the probability density function of background output which is acquired by the original data in the qth subset. Thus, a smaller $\hat{f}_q(\delta_q(\mathbf{y}_q(k)))$ represents that $(\delta_q(\mathbf{y}_q(k)))$ is closer to the tail of $\hat{f}_q$, which indicates that the kth pixel is probably belong to target. Base on the hypothesis above, the fuzzy support function $h_q(\delta_q(\mathbf{y}_q(k))) : \delta_q(\cdot) \rightarrow [0, 1]$ in the q-th subset is defined as:

$$h_q(\delta_q(\mathbf{y}_q(k))) = 1 - \frac{\hat{f}_q(\delta_q(\mathbf{y}_q(k)))}{\sum_{k=1}^{N}\hat{f}_q(\delta_q(\mathbf{y}_q(k)))} \tag{5.74}$$

The extent of detecting sample belongs to the target under the support of the q-th subset. Here, piecewise function is used to show the completed nonparametric fuzzy support function of Q band subset:

$$h(x_q(\mathbf{y}_q(k))) = \begin{cases} h_1(\delta_1(\cdot)) & q = 1 \\ h_2(\delta_2(\cdot)) & q = 2 \\ \vdots & \vdots \\ h_Q(\delta_Q(\cdot)) & q = Q \end{cases} \tag{5.75}$$

5.4.4 Fuzzy Density Estimation

The fuzzy density g_q should be estimated to acquire the confidence of each band subsets. The data in the qth subset is described as follows base on the given model in (5.1):

$$\mathbf{X}_q = \mathbf{S}_q + \mathbf{W}_q = \mathbf{B}_q + \mathbf{T}_q + \mathbf{W}_q \tag{5.76}$$

Due to the high similarity among the bands in one subset, the noise level in one subset are assumed to be the same while different noise levels appear in different subsets. The noise is assumed to be independent identical distributions while the noise relativity among bands should be ignored or set zero. Signals and noises are statistically uncorrelated. The covariance matrix of the q-th subset is:

$$
\begin{aligned}
\mathbf{R}_{\mathbf{X}_q} &= E[(\mathbf{X}_q - E\mathbf{X}_q)^H (\mathbf{X}_q - E\mathbf{X}_q)] \\
&= E[(\mathbf{S}_q - E\mathbf{S}_q + \mathbf{W}_q)^H (\mathbf{S}_q - E\mathbf{S}_q + \mathbf{W}_q)] \\
&= E[(\mathbf{S}_q - E\mathbf{S}_q)^H (\mathbf{S}_q - E\mathbf{S}_q)] + E[(\mathbf{W}_q^H \mathbf{W}_q)]
\end{aligned}
\tag{5.77}
$$

where $E[(\mathbf{W}_q^H \mathbf{W}_q)] = \sigma_{\mathbf{N}q}^2 \mathbf{I}_{lq \times lq}$. Diagonalize the signal part, we can get $\mathbf{R}_S = E[(\mathbf{S}_q - E\mathbf{S}_q)^H (\mathbf{S}_q - E\mathbf{S}_q)] = \mathbf{U}_{\mathbf{S}_q} \mathbf{\Lambda}_{\mathbf{S}_q} \mathbf{U}_{\mathbf{S}_q}^H$ where $\mathbf{\Lambda}_{\mathbf{S}_q} = diag(\sigma_{(1,Sq)}^2, \sigma_{(2,Sq)}^2, \cdots, \sigma_{(r,Sq)}^2, 0, 0, \ldots 0)$ is the $r_{\mathbf{S}_q}$ ordered diagonal matrix with nonzero eigenvalue $\sigma_{(1,Sq)}^2 \geq \sigma_{(2,Sq)}^2 \geq \cdots \geq \sigma_{(r,Sq)}^2$. These eigenvalues contain background and target energy. Additional Gaussian white noise only increases the latitude of noise covariance $\sigma_{\mathbf{N}q}^2$. Thus, $\mathbf{R}_{\mathbf{X}_q}$ is full rank which has $l_q = r_{\mathbf{S}_q}$ eigenvalues equal to $\sigma_{\mathbf{N}q}^2$ in the ideal conditions.

$$
\begin{aligned}
\mathbf{R}_{\mathbf{Y}_q} &= \mathbf{U}_{\mathbf{S}_q} \mathbf{\Lambda}_{\mathbf{S}_q} \mathbf{U}_{\mathbf{S}_q}^H + \sigma_{\mathbf{N}q}^2 \mathbf{I}_{lq \times lq} \\
&= \mathbf{U}_{\mathbf{S}_q} (\mathbf{\Lambda}_{\mathbf{S}_q} + \sigma_{\mathbf{N}q}^2 \mathbf{I}_{lq \times lq}) \mathbf{U}_{\mathbf{S}_q}^H \\
&= \mathbf{U}_{\mathbf{S}_q} \mathbf{\Lambda}_{\mathbf{Y}_q} \mathbf{U}_{\mathbf{S}_q}^H
\end{aligned}
\tag{5.78}
$$

where $\mathbf{\Lambda}_{\mathbf{Y}_q} = diag(\sigma_{(1,Sq)}^2 + \sigma_{\mathbf{N}q}^2, \sigma_{(2,Sq)}^2 + \sigma_{\mathbf{N}q}^2, \ldots, \sigma_{(r,Sq)}^2 + \sigma_{\mathbf{N}q}^2, \ldots, \sigma_{\mathbf{N}q}^2)$. Assuming that $r_{\mathbf{B}_q}$ is the order of background signal subspace, $r_{\mathbf{T}_q}$ is the order of target subspace. The target and background signal are linear independent and integrally composed the signal subspace, so $r_{\mathbf{B}_q} + r_{\mathbf{T}_q} = r_{\mathbf{S}_q}$. Target energy is much less than background energy due to the sparsity of target. So the eigenvalues of target and background are assumed to satisfy with $(\sigma_{(\cdot,Bq)}^2) \geq \max(\sigma_{(\cdot,Tq)}^2)$, where $\sigma_{(\cdot,Bq)}^2$ and $\sigma_{(\cdot,Tq)}^2$ are the eigenvalues of target and background. So $\mathbf{\Lambda}_{\mathbf{S}_q}$ is rewritten as $\mathbf{\Lambda}_{\mathbf{S}_q} = diag(\sigma_{(1,Bq)}^2 + \sigma_{(2,Bq)}^2, \ldots, \sigma_{(r_{Bq},Bq)}^2, \ldots, \sigma_{(1,Tq)}^2, \sigma_{(2,Tq)}^2, \ldots, \sigma_{(r_{Tq},Tq)}^2, 0, \ldots, 0$ and $\mathbf{R}_{\mathbf{Y}_q}$ is denoted as

$$
\begin{aligned}
\mathbf{R}_{\mathbf{Y}q} &= [\mathbf{U}_{\mathbf{B}q} \mathbf{U}_{\mathbf{T}q} \mathbf{U}_{\mathbf{N}q}] \begin{bmatrix} \mathbf{\Sigma}_{\mathbf{B}q} & & \\ & \mathbf{\Sigma}_{\mathbf{T}q} & \\ & & \mathbf{\Sigma}_{\mathbf{N}q} \end{bmatrix} \begin{bmatrix} \mathbf{U}_{\mathbf{B}q}^H \\ \mathbf{U}_{\mathbf{T}q}^H \\ \mathbf{U}_{\mathbf{N}q}^H \end{bmatrix} \\
&= \mathbf{U}_{\mathbf{B}q} \mathbf{\Sigma}_{\mathbf{B}q} \mathbf{U}_{\mathbf{B}q}^H + \mathbf{U}_{\mathbf{T}q} \mathbf{\Sigma}_{\mathbf{T}q} \mathbf{U}_{\mathbf{T}q}^H + \mathbf{U}_{\mathbf{N}q} \mathbf{\Sigma}_{\mathbf{N}q} \mathbf{U}_{\mathbf{N}q}^H
\end{aligned}
\tag{5.79}
$$

where $\mathbf{U}_{\mathbf{B}q}$ is the $l_q \times r_{\mathbf{B}q}$ matrix whose column vectors are corresponding to the $r_{\mathbf{B}q}$ eigenvectors of background signals. $\mathbf{U}_{\mathbf{T}q}$ is the $l_q \times r_{\mathbf{T}q}$ matrix whose column vectors are corresponding to the $r_{\mathbf{T}q}$ eigenvectors of target signals. $\mathbf{U}_{\mathbf{N}q}$ is the rest $l_q - r_{\mathbf{S}q}$ eigenvectors that corresponds to noise. $\mathbf{\Sigma}_{\mathbf{B}q}, \mathbf{\Sigma}_{\mathbf{T}q}$ and $\mathbf{\Sigma}_{\mathbf{N}q}$ are denoted as the following diagonal matrix:

$$\boldsymbol{\Sigma}_{\mathbf{B}q} = diag(\sigma^2_{(1,\mathbf{B}q)} + \sigma^2_{\mathbf{N}q}, \sigma^2_{(2,\mathbf{B}q)} + \sigma^2_{\mathbf{N}q}, \ldots, \sigma^2_{(r_{Bq},\mathbf{B}q)} + \sigma^2_{\mathbf{N}q}) \qquad (5.80)$$

$$\boldsymbol{\Sigma}_{\mathbf{T}q} = diag(\sigma^2_{(1,\mathbf{T}q)} + \sigma^2_{\mathbf{N}q}, \sigma^2_{(2,\mathbf{T}q)} + \sigma^2_{\mathbf{N}q}, \ldots, \sigma^2_{(r_{Tq},\mathbf{T}q)} + \sigma^2_{\mathbf{N}q}) \qquad (5.81)$$

$$\boldsymbol{\Sigma}_{\mathbf{N}q} = diag(\sigma^2_{\mathbf{N}q}, \sigma^2_{\mathbf{N}q}, \ldots, \sigma^2_{\mathbf{N}q}) \qquad (5.82)$$

Considering the effect caused by SNR, fuzzy density g_q is defined as the ratio of target signals energy and noise signals energy:

$$g_q = \frac{\sum_{i=1}^{r_{Tq}} \sigma^2_{(i,\mathbf{T}q)}}{l_q \times \sigma^2_{\mathbf{N}q}} \qquad (5.83)$$

Due to the limited amount of samples, the eigenvalues have a monotone decreasing distribution in practical [3]. To acquire better estimation of $\sigma^2_{\mathbf{N}q}$, let $\hat{\sigma}^2_{\mathbf{N}q} = \frac{1}{l_q - r_{Sq}} \sum_{i=r_{Sq}+1}^{l_q} \sigma^2_{(i,q)}$, where $\sigma^2_{(i,q)}$ represents l_q eigenvalues of $\mathbf{R}_{\mathbf{Y}q}$ reduced in order, namely $\sigma^2_{(1,q)} \geq \sigma^2_{(2,q)} \geq \cdots \geq \sigma^2_{(l_q,q)}$, so (5.83) is changed:

$$g'_q = \frac{\sum_{i=1}^{r_{Tq}} \hat{\sigma}^2_{(i,\mathbf{T}q)}}{l_q \times \hat{\sigma}^2_{\mathbf{N}q}} = \frac{\sum_{i=r_{Bq}+1}^{r_{Sq}} (\sigma^2_{(i,q)} - \hat{\sigma}^2_{\mathbf{N}q})}{l_q \times \hat{\sigma}^2_{\mathbf{N}q}} \qquad (5.84)$$

$r_{\mathbf{B}q}$ and $r_{\mathbf{T}q}$ should be estimated because the orders of signal and target subspace are unknown. The definition of $r_{\mathbf{B}q}$ is:

$$r_{\mathbf{B}q} = \min(\underset{t}{\arg}(\frac{\sum_{i=1}^{t} \sigma^2_{(i,q)}}{\sum_{i=1}^{l_q} \sigma^2_{(i,q)}} \geq \Delta_{\mathbf{B}q})) \quad (t \leq l_q) \qquad (5.85)$$

$\Delta_{\mathbf{B}q}$ is the minimal accumulated variance contribution rate of background variance.

To get $r_{\mathbf{T}q}$, eigenvalues $\sigma^2_{(1,q)}, \sigma^2_{(2,q)}, \ldots, \sigma^2_{(l_q-1,q)}$ are firstly calculated:

$$\alpha_{(i,q)} = \frac{\sigma^2_{(i,q)} - \sigma^2_{(i+1,q)}}{\sigma^2_{(i,q)}}, i = 1, 2, \ldots, l_q - 1 \qquad (5.86)$$

The difference of eigenvalue changing ratio is:

$$\beta_{(i,q)} = \alpha_{(i+1,q)} - \alpha_{(i,q)} \qquad (5.87)$$

Seek the i_q^* eigenvalue satisfying with the following conditions by $\alpha_{(\cdot,q)}$ and $\beta_{(\cdot,q)}$.

(1) *Selecting Rule*: let $\Omega_q(\rho)$ be the set which contains the ρ maximal ratios in subsequence $\alpha_{(r_{B_q}+1,q)}$, $\alpha_{(r_{B_q}+2,q)}, \ldots, \alpha_{(l_q-1,q)}$. $\Omega_q(\rho)$ is the index set of the ratios in $\Phi_q(\rho)$, so we can acquire $i_q' \in \Phi_q(\rho)$ and $\alpha_{(i_q',q)} \in \Omega_q(\rho)$.

(2) *Minimum Rule*: Assume $i_q^* = \arg(\max\limits_{i \ i\in\Phi_q(\rho)}(\beta_{(i,q)}))$. $\beta_{(i,q)}$ describes the changes of $\alpha_{(i+1,q)}$ and $\alpha_{(i,q)}$. $\alpha_{(i,q)}$ represents the descend ratio of the ith eigenvalue to the $(i+1)$th eigenvalue. Thus, the most sharp change of $\alpha_{(i,q)}$ is at i_q^*, which is the inflection point of target eigenvalue and noise eigenvalue, namely $i_q^* - r_{s_q}$. As we have $r_{B_q} + r_{T_q} = r_{s_q}$,

$$r_{Tq} + i_q^* - r_{Bq} \tag{5.88}$$

5.4.5 Experiment Results

The LCTF imaging spectro-polarimeter generates 30 bands across the whole spectral range 0.4–0.7 μm. A multi-band polarization image is acquired by the imaging spectro-polarimeter, which is located on top of a large building. The scene is composed of grassland and four small objects located on the grassland in different locations. (These four objects, which are painted in green, are hard to separate from the green background in the panchromatic band.) The panchromatic image of the whole scene and the location of the targets are shown in Fig. 5.3.

Detection results can be acquired using subset fuzzy integral fusion detection method in this section. Detailed results are shown in Fig. 5.4 where (a) is the result using RX method and (b) is the result using subset fuzzy integral fusion detection method.

Fig. 5.3 Panchromatic image of experimental scene

(a)   **(b)**

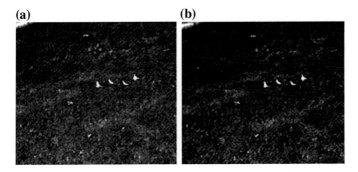

Fig. 5.4 Results of multi-band polarization image detection. **a** The result of RX method. **b** The result of subset fuzzy integral fusion method

5.5 Multi-detector Maximum Entropy Fusion Hypothesis Detection

5.5.1 Single Likelihood Hypothesis Detection Model

Target detection can be simply considered as a dichotomy problem that every pixel is usually recognized as either target or background by binary hypothesis test (H_0: no target (background), H_1: target). The nonstructural model is used in multi-band polarization imagery analysis to model the background information. The background is assumed to obey a certain distribution, such as the common used Gaussian distribution, mixing multivariate Gaussian distribution, elliptically contoured distribution [22–26]. A more general probability density function $p_0(\cdot; \theta_0)$ is used to describe background, where θ_0 is the parameter estimated from samples. In a similar way, target samples are assumed to be described by a high dimension distribution $p_1(\mathbf{x}; \theta_1)$, where θ_1 is the corresponding parameter. Let $p_0(\cdot; \theta_0)$ and $p_1(\mathbf{x}; \theta_1)$ denote the background samples and target samples, so there is a hypothesis model:

$$H_0 : \mathbf{x} \sim p_0(\mathbf{x}; \theta_0)$$
$$H_1 : \mathbf{x} \sim p_1(\mathbf{x}; \theta_1)$$

(5.89)

In hypothesis problems, due to the low probability characteristic, target may be judged as background using the Bayes criteria. Thus, Neyman-Pearson criterion should be used to guarantee the maximal detection probability with a certain false alarm ratio. Optimal decision can be acquired by likelihood ratio rest:

$$
\begin{aligned}
\Lambda(\mathbf{x}) &= \frac{L_1(\mathbf{x})}{L_0(\mathbf{x})} \\
&= \frac{p(\mathbf{x}|H_1)}{p(\mathbf{x}|H_0)} \\
&= \frac{p_1(\mathbf{x}; \theta_1)}{p_0(\mathbf{x}; \theta_0)} \begin{cases} \geq \eta & \text{target} \\ < \eta & \text{background} \end{cases}
\end{aligned}
\tag{5.90}
$$

where η is the detection threshold. Parameters θ_0 and θ_1 can be acquired using the maximum likelihood estimation without prior information. At this time, the likelihood test changes to a generalized likelihood ratio test. There is no prior information about multi-band polarization characteristics and distribution of target in hypothesis test. The signal clutter additional model should be converted to the target-background additional model. Nonstructural background information should only be contained in H_0, while both nonstructural background and certain target information should only be contained H_1. The hypothesis test model is rewritten as:

$$
\begin{aligned}
H_0 &: \mathbf{x} \sim p_0(\cdot; \theta_0) & \text{no target} \\
H_1 &: (\mathbf{x} - \mathbf{s}) \sim p_0(\cdot; \theta_0) & \text{target}
\end{aligned}
\tag{5.91}
$$

where $\mathbf{s}$ is the certain target information. So generalized likelihood ratio test is simplified as a single likelihood test of sample $\mathbf{x}$ under zero hypothesis, the detector is as follows [4]:

$$
\delta(\mathbf{x}) \equiv -\log(p_0(\mathbf{x}; \theta_0)) \underset{H_0}{\overset{H_1}{\gtrless}} \eta
\tag{5.92}
$$

where η is the detection threshold.

5.5.2   Kernel Density Estimation

Outputs from each detector are considered as information sources which should be fused by the maximum entropy criteria, so the probability density should be acquired. However, it is difficult to describe multi-band polarization imagery with fixed parametric statistical model due to the differences between parametric statistical model and real data [10–14]. So the modeling methods always coming with model error that seriously affect detection results. For example, the real multi-band polarization image data always has the long tail phenomenon [14] which will make the Gaussian distribution hypothesis of original data not valid, so the Chi-square cannot be used to describe the nine outputs of quadratic detector δ_{RXD}^C, δ_{RXD}^R, δ_{UTD}^C, δ_{UTD}^R, δ_{MRXD}^C, δ_{MRXD}^R, δ_{NRXD}^C, δ_{NRXD}^R and δ_{RX-UTD}^C.

Adaptive kernel density estimation [10], which is without modeling and estimating the model parameters, can overcome the above drawbacks. Adaptive kernel density estimation is a two-step kernel density estimator. After estimating samples in the pre-estimator, the samples are used to adjust the pre-estimator to obtain improved results. Moreover, the abnormalities are considered owing low probability and mainly distribute in the tail of measure probability density. The adaptive kernel estimation can adaptively adjust the width of kernel function using local width factor according to the sample distribution. It can finely describe the local tiny variations and remain the long tail feature to estimate the probability more accurately.

Let $\delta_m(n)$ be the output of the m-th ($m \in \{1, 2, \ldots, M\}$) detector at the pixel n ($n \in \{1, 2, \ldots, N\}$). The probability density estimation of all samples output by the m-th detector is shown:

(1) Choose a kernel density estimator with fixed window width, the m-th probability density estimation, which is acquired from whole of N samples, is denoted as $\tilde{f}_m(t)$. An arbitrary sample should satisfy with
(2) Calculate the local window width factor $\lambda_m(n)$

$$\lambda_m(n) = \left\{ \tilde{f}_m(\delta_m(n)) / \exp(N^{-1} \sum \tilde{f}_m(\delta_m(n))) \right\}^{-\alpha} \tag{5.93}$$

where $0 \le \alpha \le 1$, α is the sensitivity parameter.
(3) The adaptive kernel density estimator is $\tilde{f}_m(t)$:

$$f_m(t) = N^{-1} \sum h^{-1} \lambda_m^{-1}(n) K\left\{ h^{-1} \lambda_m^{-1}(n)(t - \delta_m(n)) \right\} \tag{5.94}$$

K is a kernel function which is generally in Gaussian form $K(x) = (2\pi)^{-1/2} \exp(-\frac{x^2}{2})$, α is set to 0.5 [13]. As the choice of kernel density estimator has little effect on the final results, the standard kernel estimator $\tilde{f}_m(t) = (1/Nh) \sum_{n=1}^{N} K((t - \delta_m(n))/h)$ with a fixed window width is used as a pre-estimator. h is the width of kernel function which can be determined by (ROT) criterion [8, 9]. The window width of pre-estimator and final estimator should be the same to attain the best estimation performance [9].

5.5.3 Detector Fusion Based on Maximum Entropy

There are many methods for multiple detectors fusion such as the logical fusion and relative fusion. In this text, the maximum entropy criterion is used as fusion rule to avoid subjective assumptions [4, 21]. The maximum entropy model can estimate probability characteristics of random variables with the remaining of the maximum information amount. It can also make fusion result the best under the definition of information entropy without any subjective assumptions. So the distribution acquired by maximum entropy criterion contains the maximum information and objective degree.

Consider all of samples $\mathbf{x}(n)n \in \{1, 2, \ldots, N\}$ in multi-band polarization image data as a set of events, y is the decision output which come from binary hypothesis test set $\{H_0$: no target, H_1: target$\}$ in detection problems. The effect on output probability density function can be ignored as the sparsity of samples. The output probability density function $\hat{f}_m(t)$ of the m-th detector is approximately seen as the probability density function after the original data passing through the m-th detector. We can calculate the probability density of every data sample $\mathbf{x}(n)$ in the condition H_0:

$$p_{m0}(\mathbf{x}(n)|y : H_0, m) = \hat{f}_m(\delta_m(n)) \tag{5.95}$$

$d_m(n) = p_{m0}(\mathbf{x}(n)|y : H_0 m)$, $d_m(n)$ is the measure of the given sample $\mathbf{x}(n)$. Here, we try to acquire the jointly conditional probability density to the given M detector in the condition H_0:

$$p_J(\mathbf{x}(n)|d_1(n), d_2(n), \ldots, d_M(n)) \tag{5.96}$$

Entropy is the function of probability density $p(\mathbf{x})$, is independent with the value of $\mathbf{x}$:

$$H_P(p) = - \int_\Theta p(\mathbf{x}) \log_a p(\mathbf{x}) d\mathbf{x} \tag{5.97}$$

Estimating the probability density with the maximum entropy criterion is to find a probability density distribution which can achieve the maximal value of entropy.

$$\hat{p}(\mathbf{x}) = \arg(\max_p(H_p(p))) \tag{5.98}$$

The function in (5.98) should satisfy with following constraints:

$$\begin{aligned} &\hat{p}_1(\mathbf{x}) \geq 0 \\ &\int_\Theta \hat{p}_1(\mathbf{x})d\mathbf{x} = 1 \\ &\int_\Theta \hat{p}_1(\mathbf{x})r_i(\mathbf{x})d\mathbf{x} = \alpha_i, \quad 1 \leq i \leq m \end{aligned} \tag{5.99}$$

where $\alpha_i(1 \leq i \leq m)$ are the moments of the probability density estimation, $r_i(\mathbf{x})$ is the expression of the moments of sample $\mathbf{x}$. The unique jointly probability density function can be acquired by fusing the output probability density functions output of every detector. The unique jointly probability density function, which contains the maximum uncertainty, remains the maximum amount of information and does not affect by any subjective factors except detector outputs.

The maximum entropy estimation of $p_J(\mathbf{x}(n)|\mathbf{d})$ is:

$$\hat{p}_J(\mathbf{x}(n)|\mathbf{d}) = \arg(\max H(p_J(\mathbf{x}(n)|\mathbf{d})))$$
$$= \arg(\max(-\sum_n p_J(\mathbf{x}(n)|\mathbf{d}) \log p_J(\mathbf{x}(n)|d))) \qquad (5.100)$$

where $\mathbf{d} = [d_1(\cdot), d_2(\cdot), \ldots, d_M(\cdot)]^T$, H is entropy of p_J.

Theorem 5.3 *Normal probability density function makes the entropy H_f the maximum in all of probability density functions f with given covariance matrix $\mathbf{\Sigma}$. The maximum entropy is:*

$$\max H_f = \frac{1}{2}\log(2\pi e)^n \det|\mathbf{\Sigma}| \qquad (5.101)$$

Let D_M be the distribution space with standard normal edge distribution in R^M. $D_M(\mathbf{C})$ is a family of distribution with correlation matrix $\mathbf{C}$ in D_M. The multivariate normal distribution $N(0, \mathbf{C})$ has the maximum entropy in $D_M(\mathbf{C})$. Firstly the output of the m-th detector should be projected to the standard Gaussian space by projection transformation T_m to make the data after transformation obey with Gaussian distribution:

$$z_m(n) = T_m(\delta_m(n)) = \Phi^{-1}(\Psi_m(\delta_m(n))) \qquad (5.102)$$

Ψ_m is the accumulated experienced probability density function of the m-th detector, Φ^{-1} is the inverse function of Ψ_m. Then the multivariate normal distribution model should be constructed in $D_M(\mathbf{C})$ to acquire the jointly probability density function with the maximum entropy in projection space:

$$\hat{q}_0(\mathbf{Z}(n)) = (2\pi)^{-1}\|\mathbf{C}\|^{-1}\exp(-\frac{1}{2}\mathbf{Z}(n)^T\mathbf{C}\mathbf{Z}(n)) \qquad (5.103)$$

where $\mathbf{Z}(n) = [z_1(n), z_2(n), \ldots, z_m(n)]^T$, $\hat{\mathbf{C}} = \mathbf{Z}\mathbf{Z}^T$ is the correlation matrix. $\hat{q}_0(\mathbf{Z}(n))$ is the jointly probability density estimation of sample $\mathbf{x}(n)$. So the jointly probability density estimation $\hat{p}_J(\mathbf{x}(n)|\mathbf{d})$ in original space can be acquired using mapping $T(\delta_1(n), \delta_2(n), \ldots, \delta_M(n)) = (z_1(n), z_2(n), \ldots, z_M(n))$ and (5.100):

$$\hat{p}_J(\mathbf{x}(n)|\mathbf{d}) = \hat{p}_J(\delta_1(n), \delta_2(n), \ldots, \delta_M(n))$$
$$= \hat{q}_0(\mathbf{Z}(n))\|\nabla T(\delta_1(n), \delta_2(n), \ldots, \delta_M(n))\|$$
$$= \|\hat{\mathbf{C}}\|^{-1}\exp\left(-\frac{1}{2}T(\delta_1(n), \delta_2(n), \ldots, \delta_M(n)^T)\right) \qquad (5.104)$$
$$(\hat{\mathbf{C}} - \mathbf{I})^{-1}T(\delta_1(n), \delta_2(n), \ldots, \delta_M(n)))\prod_{m=1}^{M} p_{m0}(\mathbf{x}(n))$$

The results of (5.100) also have the maximum entropy in the original space, so the fusion results of multiple detectors can be acquired [7]. Replace $p_0(\cdot;\theta_0)$ in (5.92) with $\hat{p}_J(\mathbf{x}(n)|\mathbf{d})$, the maximum entropy fusion detector can be acquired:

$$\Lambda(\mathbf{x}) = -\log(\hat{p}_J(\mathbf{x}(\cdot)|\mathbf{d})) \underset{H_0}{\overset{H_1}{\gtrless}} \eta \qquad (5.105)$$

where η denotes the detecting threshold.

5.5.4   Experiment Results

The LCTF imaging multi-band polarimeter generates 30 bands across the whole spectral range 0.4–0.7 μm, The multi-band polarization imagery is acquired by the imaging multi-band polarimeter, which is located on top of a large building. The scene is composed of grassland and six small objects located on the grassland in different locations. (These six objects, which are painted green, are hard to separate from the green background in the panchromatic band.) The panchromatic image of the whole scene and the location of the targets are shown in Fig. 5.5.

Detection results can be acquired using above detection method in this section. Detailed results are shown in Fig. 5.6a is the result using RX method with threshold processing (b) is the result using subset fuzzy integral fusion detection method with threshold processing (c) is the result using the multiple detectors maximum entropy anomaly detection method with threshold processing.

Fig. 5.5 Panchromatic image of experimental scene

(a)   **(b)**

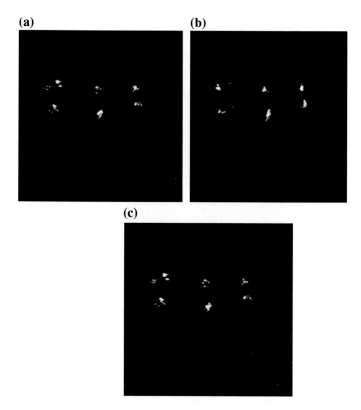

(c)

Fig. 5.6 Detection results of multi-band polarization image. **a** Result of RX method. **b** Result of subset fuzzy integral fusion detection method. **c** Result of multiple maximum entropy anomaly detection method

5.6 Fusion Detection Based on Polarization and Spectral Features

The detection methods introduced in Sects. 5.3–5.5 are available to multivariate data. Except from these methods, we can also divide the multi-band polarization image data into groups with different physical meanings according to the differences between polarized and spectral features, then use the methods in Sects. 5.3–5.5 to do the primarily identification and the results will be finally fused to achieve the final detection results.

5.6.1 Fusion Detection Algorithms

One multi-band polarization data set can be divided into two subsets: polarized spectrum subset and polarized data subset. Spectral differences are used to identify

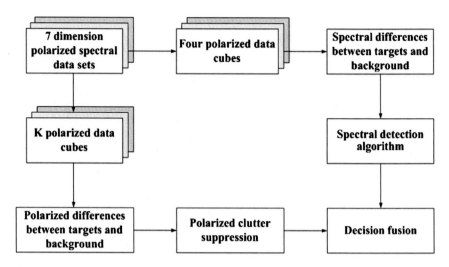

Fig. 5.7 Processing graph of multi-band polarization target detection

the target and background in polarized spectrum subset, while in polarized data subset, the signal differences are indicated by the polarized differences, such as the differences of artificial objects and natural background, and the differences of different surface materials. The physical principles in these two subsets are different so the detection results have different confidence or even be against with each other are also different. Thus, the detection results of these two subsets should be fused to attain higher accuracy and reliability. The detailed algorithm is shown as follows:

(1) Divide multi-band polarization dataset into two subsets: polarized spectrum subset and polarized data subset.
(2) Calculate spectral feature images using the detection algorithm in Sects. 5.3–5.5.
(3) Use PCA, manifold learning, wavelet analysis or other methods to choose bands, then extract polarized features images using anomaly detection method.
(4) Choose proper fusion method to fuse the polarized images and spectral feature images.

According the fusion algorithm above, the framework of the target detection for multi-band polarization images is shown in Fig. 5.7. The spectral detection methods can be the common anomaly detection methods, feature layer target matching detection method or the methods mentioned in Sect. 5.4. Different results can be acquired by combing different spectral detection algorithms. In the aspect of cluttering suppression, the polarized differences between target and background are used to clutter background signals.

The common anomaly detection methods are not able to efficiently utilize the polarized structural features. However, polarized parameter images always contain surface features, such as roughness and surface components, which can be used to identify artificial targets, natural targets, metal surface and insulator surface to improve detection performance in clutter suppression. Several methods were proposed to suppress background signals by the physical features of polarization parameters [2, 27].

(1) Normalized the Stokes images;
(2) Do wavelet decomposition on Stokes parameter images and DoLP image, so the decomposition coefficients of point (x, y) in the k-th layer at the l direction are $S_0(x, y, k, l)$, $S_1(x, y, k, l)$, $S_2(x, y, k, l)$ and $\mathrm{DoLP}(x, y, k, l)$;
(3) Set the size of feature region $M \times N$, determine values of M and N by the relative size of target and background (generally set 3×3 or 5×5);
(4) Calculate the energy coefficients of the wavelet decomposition coefficient $S_0(x, y, k, l)$, $S_1(x, y, k, l)$, $S_2(x, y, k, l)$ and $\mathrm{DoLP}(x, y, k, l)$ according to the wavelet coefficient energy in each window:

$$E_{S_0} = \frac{4}{M \times N} \sum_{x=1}^{M/2} \sum_{y=1}^{N/2} S_0(x, y, k, l)^2 \tag{5.106}$$

$$E_{S_1} = \frac{4}{M \times N} \sum_{x=1}^{M/2} \sum_{y=1}^{N/2} S_1(x, y, k, l)^2 \tag{5.107}$$

$$E_{S_2} = \frac{4}{M \times N} \sum_{x=1}^{M/2} \sum_{y=1}^{N/2} S_2(x, y, k, l)^2 \tag{5.108}$$

$$E_{DoLP} = \frac{4}{M \times N} \sum_{x=1}^{M/2} \sum_{y=1}^{N/2} DoLP(x, y, k, l)^2 \tag{5.109}$$

Considering different DoLP between artificial target and natural background, the energy coefficients E_{S_0}, E_{S_1} and E_{S_2} should be adjusted to acquire the corresponding weighted coefficients.

$$a_{S_0}(x, y, k, l) = E_{S_0} \times Do\tilde{L}P(x, y, k, l) \tag{5.110}$$

$$a_{S_1}(x, y, k, l) = E_{S_1} \times Do\tilde{L}P(x, y, k, l) \tag{5.111}$$

$$a_{S_2}(x, y, k, l) = E_{S_2} \times Do\tilde{L}P(x, y, k, l) \tag{5.112}$$

$$a_{DoLP}(x,y,k,l) = E_{DoLP} \times Do\tilde{L}P(x,y,k,l) \tag{5.113}$$

where $Do\tilde{L}P(x,y,k,l)$ is the approximate coefficient of the wavelet decomposition of DoLP.

(5) Fuse the wavelet decomposition coefficient $S_0(x,y,k,l)$, $S_1(x,y,k,l)$, $S_2(x,y,k,l)$, $DoLP(x,y,k,l)$ in every window according the following weighted criterion:

$$\begin{aligned} F(x,y,k,l) = {} & a_{S_0} \times S_0(x,y,k,l) + a_{S_1} \times S_1(x,y,k,l) \\ & + a_{S_2} \times S_2(x,y,k,l) + a_{DoLP} \times DoLP(x,y,k,l) \end{aligned} \tag{5.114}$$

(6) Do inverse transformation and get the fusion result:

$$F = IDWT(F(x,y,k,l)) \tag{5.115}$$

5.6.2 Experiment Results

The LCTF imaging multi-band polarimeter generates 30 bands across the whole spectral range 0.4–0.7 μm, The multi-band polarization imagery is acquired by the imaging multi-band polarization, which is located on top of a large building. The scene is composed of grassland and two small objects located on the grassland in different locations. (These two objects, which are painted green, are hard to separate from the green background in the panchromatic band.) The panchromatic image of the whole scene and the location of the targets are shown in Fig. 5.8.

Fig. 5.8 Panchromatic image of experimental scene

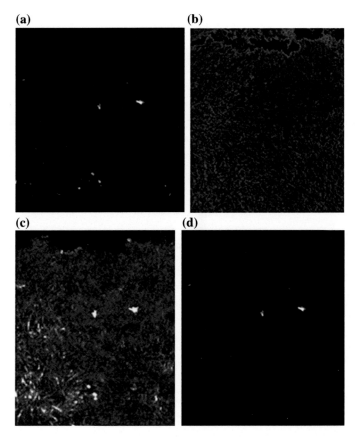

Fig. 5.9 Multi-band polarization image detection results. **a** Result of RX method. **b** Result of subset fuzzy integral fusion method. **c** Result of maximum entropy method. **d** Result of feature fusion target matching method

Detection results are shown in Fig. 5.9 where (a) is the result of RX algorithm (b) is the result of subset fuzzy integral fusion detection (c) is the result of maximum entropy detection (d) is the result of feature fusion target matching detection. Figure 5.10 shows the 3D display images of Fig. 5.9, which can absolutely represent the differences between target and background. While Fig. 5.11 shows the separation of target and background where green stripes and black stripes respectively represent the statistical distribution of target and background.

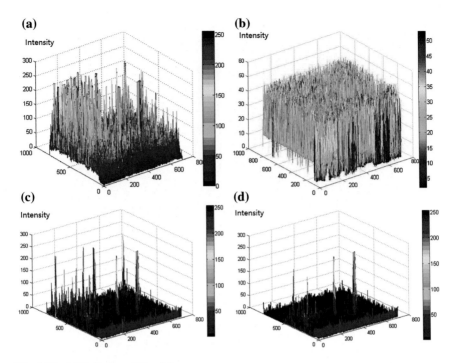

Fig. 5.10 A 3D display of Fig. 5.9

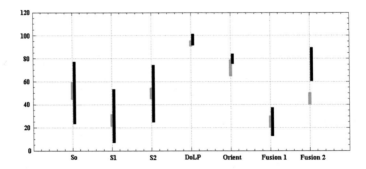

Fig. 5.11 Separation of target and background

5.7   Fusion Detection Based on DoLP Modulation

Polarization contains more surface information than intensity, such as surface roughness and conductive characteristics. Different polarized parameters S_0, *DoLP* and *Orient* reflect different features in scene. For example, S_0 reflects the geometric characteristics and structural texture. *DoLP* reflects the conductive characteristics

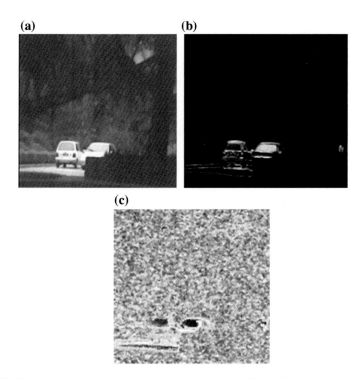

Fig. 5.12 Polarized parameter images of outdoor scenes. **a** s_0. **b** *DoLP*. **c** *Orient*

and surface roughness. *Orient* reflects the direction of surface units. If these information can be efficiently fused, the structure of target will be finely described to improve the detection performance. There are few achievements on the fusion of S_0, *DoLP* and *Orient*, A parameter image fusion method was proposed in [28] to identify target and background using *DoLP* modulation which is shown as follows.

5.7.1 DoLP Modulation Fusion

Figure 5.12 shows a group of polarized parameter images which contain car, cement pavement and plants. The scenes can be divided into natural scenes and artificial scenes. In the field of remote sensing, artificial targets are considered owing smooth surface and higher *DoLP* while natural scenes are assumed to have rough surface with smaller *DoLP*. As we can see from Fig. 5.12, the artificial targets, such as the road and cars, have higher *DoLP*. Moreover, the light reflected by surface of conductors has relatively smaller *DoLP* while the light reflected by surface of insulators has relatively higher *DoLP*.

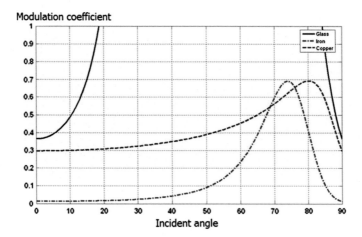

Fig. 5.13 Relationship of different *DoLP* modulation coefficient and incident angle

Therefore, due to the identification advantages of *DoLP*, a *DoLP* modulation coefficient can be designed to enlarge the differences of *DoLP* between artificial targets and natural targets in target detection. The designed *DoLP* modulation coefficient should be continued and monotone increasing as follows:

$$M_{DoLP} = DoLP \cdot \log(1 + DoLP) \tag{5.116}$$

Figure 5.13 shows how the polarized angle varied from incident angle of different metals (iron, copper) and nonmetal (glass). *Orient* is presented to change slowly in the local smooth areas and changes sharply in the rough areas. Being similar with the polarized modulation coefficient, *Orient* modulation coefficient can also be designed to enlarge the differences of *Orient* between smooth surface and rough surface.

$$\overline{Orient} = \sqrt{\frac{M'N'}{\sum_{x=1}^{M'} \sum_{y=1}^{N'} (Orient(x, y) - \mu)^2}} \tag{5.117}$$

where μ is the average of local area $M' \times N'$. Pixel values in smooth areas are higher than those in rough areas. Fusion algorithm can be described as follows by fusing S_0, M_{DoLP} and $\overline{Orient}$:

(1) Pre-process images S_0, M_{DoLP} and $\overline{Orient}$
(2) Seek the public information shared by S_0, M_{DoLP} and $\overline{Orient}$:

$$C_o = S_0 \cap DoLP \cap \overline{Orient} \tag{5.118}$$

$$A \cap B(i,j) = \min\{A(i,j), B(i,j)\}.$$

(3) Calculate the unique information in each image

$$\begin{cases} DoLP^* = DoLP - Co \\ Orient^* = \overline{Orient} - Co \\ S_0^* = S_0 - Co \end{cases} \tag{5.119}$$

(4) Adjust images using the unique information acquired in step (3)

$$\begin{cases} DoLP^{**} = DoLP - Orient^* - S_0^* \\ Orient^{**} = \overline{Orient} - DoLP^* - S_0^* \\ S_0^{**} = S_0 - DoLP^* - Orient^* \end{cases} \tag{5.120}$$

(5) Synthesize a false color image using result in step (4)

$$F = RGB(S_0^{**}, DoLP^{**}, Orient^{**}) \tag{5.121}$$

(6) Adjust the false color image using *DoLP* modulation coefficient

$$Fusion = M_{DoLP} \cdot F \tag{5.122}$$

The fusion result of Fig. 5.12 is shown in Fig. 5.14, artificial targets and natural scene are denoted as different colors to efficiently distinguish from each other.

Fig. 5.14 Fusion result of polarized parameter image

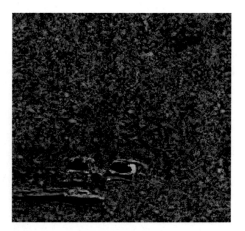

5.7.2   False Color Coding Fusion of Spectral and Polarization Images

As shown in Fig. 5.14, polarized parameters S_0, *DoLP* and *Orient* can be mapped to the Intensity, Hue and Saturation spaces through false color. While spectral images can also be described in the IHS space, so the false color coding fusion of spectral and polarized images can be achieved as follows:

(1)   Transfer the spectral image to IHS color space;
(2)   Map the polarized parameters into IHS space according to Fig. 5.15.
(3)   Fuse the intensity, hue, saturation in IHS space. Then the results are inversely mapped as S_0, *DoLP* and *Orient* images. Finally use the polarized modulation fusion method in Sect. 5.7.1 to fuse the polarized parameter images.

False color fusion result of Fig. 5.12 is shown in Fig. 5.16.

From the comparison of Figs. 5.14 and 5.16, the background disturbance can be efficiently suppressed by spectral and polarized information. Figure 5.18 shows the result of simulated camouflaged target whose scene is shown in Fig. 5.17. Five

Fig. 5.15 Mapping relationship of polarized parameter and intensity, hue, saturation

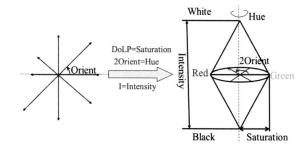

Fig. 5.16 False color fusion result of Fig. 5.12

Fig. 5.17 Camouflaged
targets scene

Fig. 5.18 False color coding
fusion result of the image in
Fig. 5.17

camouflaged aircraft model lie on the ground, and the light condition changes
sharply, the targets can be accurately detected using the given false color coding
fusion method of polarized and spectral images.

References

1. Schott JR (2007) Remote sensing—the image chain approach. Oxford University Press, New York
2. Zhao Y, Gong P, Pan Q (2008) Object detection by spectropolarimeteric imagery fusion. IEEE Trans Geosci Remote Sens 46(10):3337–3345
3. Manolakis D, Shaw G (2002) Detection algorithms for hyperspectral imaging application. IEEE Signal Process Mag 19(1):29–43
4. Li D (2004) Signal statistical detection and estimation. Science Press, Beijing, pp 26–112
5. Zhang Y, Fang K (2003) Multivariate statistical analysis. Science Press, Beijing, pp 53–95
6. Ranney KI, Soumekh M (2006) Hyperspectral anomaly detection within the signal subspace. IEEE Geosci Remote Sens Lett 3(3):312–316
7. Fang J, Yu H (1994) Information theory and coding. Electronic Industry Press, Beijing
8. Fryer MJ (1977) A review of some non-parametric methods of density estimation. J Appl Math 20:335–354
9. Devroye L, Gyorfi L (1987) A course in density estimation. Birkhäuser, Boston
10. Mugdadi AR (2004) A bandwidth selection for kernel density estimation of functions of random variables. Comput Stat Data Anal 47(11):49–62
11. Silverman BW (1986) Density estimation for statistics and data analysis. Chapman and Hall, London, pp 1–119
12. Reed IS, Yu X (1990) Adaptive multiple-band CFAR detection of an optical pattern with unknown spectral distribution. IEEE Trans Acoust Speech Signal Process 38(10):1760–1770
13. Yu X, Reed IS, Stocker AD (1993) Comparative performance analysis of adaptive multispectral detectors. IEEE Trans Signal Process 41(8):2639–2656
14. Yu X, Hoff LE, Reed IS, Chen A, Stotts LB (1997) Automatic target detection and recognition in multi-band imagery: a unified ml detection and estimation approach. IEEE Trans Image Process 6(1):143–156
15. Manolakis D (2003) Detection algorithms for hyperspectral imaging application: a signal processing perspective. In: IEEE proceedings of workshop on advances in techniques for analysis of remotely sensed data, pp 378–384
16. Lin H, Quan P, Zhao Y (2005) An automatic target detection algorithm for hyperspectral imagery based on feature-level fusion. International Society for Optics and Photonics, SAR and Multispectral Image Processing
17. Di W, Pan Q, Zhao Y (2006) Multiple-detector fusion for anomaly detection in multispectral imagery based on maximum entropy and nonparametric estimation. In: The 8th international conference on signal processing, vol 3, pp 16–20
18. He L, Pan Q, Zhao Y (2006) Source based small targets detection for hyperspectral imagery using evidential reasoning. In: International conference on machine learning and cybernetics, pp 1979–1984
19. Di W, Pan Q, He L (2008) Anomaly detection in hyperspectral imagery by fussy integral fusion of band-subsets. Photogram Eng Remote Sens 74(2):201–213
20. He L, Pan Q, Di W (2008) Anomaly detection in hyperspectral imagery by maximum entropy and nonparametric estimation. Pattern Recogn Lett 29(9):1392–1403
21. Di W, Pan Q, Zhao Y (2008) Anomaly target detection in hyperspectral imagery based on band subset fusion by fuzzy integral. J Electron Inf Technol 30(2):267–271
22. He L, Pan Q, Zhao Y (2005) Anomaly target detection in hyperspectral imagery based on band subset feature fusion. Acta Photonica Sinica 34(11):1752–1755
23. He L, Pan Q, Wei D (2007) A small-target detector based on single likelihood test for hyperspectral imagery. Acta Photonica Sinica 27(12):2155–2162
24. He L, Pan Q, Zhao Y (2007) Target detection in hyperspectral imagery based on linear mixing model reconstructed from measurements. Acta Electronica Sinica 35(1):23–27
25. Di W, Pan Q, He L (2007) Target detection in multi-channel imagery based on maximum entropy fusion in multiple detectors. Acta Photonica Sinica 36(7):1338–1344

26. He L, Pan Q, Zhao Y (2006) CFAR target detection in unknown background based on subspace projection in aerial hyperspectral imagery. Acta Aeronautica Astronautica Sinica 26 (4):657–662
27. Zhao Y, Pan Q, Zhang H (2006) Object detection by fusion spectral and polarimetric imagery. In: Proceedings SPIE remote sensing and infrared devices and systems, 2006
28. Zhao Y, Zhang L, Zhang David (2009) Object separation by polarimetric and spectral imagery fusion. Comput Vis Image Underst 113(8):855–866

Chapter 6
Bio-inspired Multi-band Polarization Imaging

Multi-band polarization imaging has a variety of applications including land cover classification, military target detection, 3D surface reconstruction, and glare/shadow removal [1]. A comprehensive utilization of spatial, spectral, and polarization information is effective to detect hidden or camouflaged targets which may not be discoverable by traditional optical imaging techniques [2]. Present multi-band polarization sensors rely on dispersion, beam split or interference to obtain spectral information, while polarization information in different direction is acquired by changing polarization angles [2]. Acquisition of multiple images in different spectral bands and polarization angles may take a relatively long time, a large amount of data, and limited field of view [1]. Conventional multi-band polarization imaging techniques may not satisfy the requirements in modern warfare in terms of dynamic environments, larger field of view, and fast response time. Recent research indicates that some aquatic organisms like dragonfly nymphs, mantis shrimps, and cuttlefish hunt with their multi-band polarization vision. Such observations sparked the studies on multi-band polarization imaging techniques [3–6].

6.1 Multi-band Polarization Vision

6.1.1 Polarization Vision

Polarization vision enables distinguishing two directions of the electric-field and measures the rate of polarization. Bio-inspired polarization vision helps compensate limitations with human vision using polarization characteristics differences between targets and background [7]. Recent research indicates that mantis shrimps, dragonfly nymphs, cephalopods and some fishes are sensitive to polarized lights [4, 8–10]. Then a group of polarization suitable special ommatidia in the dorsal rim area (DRA) of a compound eye are found [11]. Every ommatidium contains two

© National Defense Industry Press, Beijing and Springer-Verlag Berlin Heidelberg 2016   155
Y. Zhao et al., *Multi-band Polarization Imaging and Applications*,
Advances in Computer Vision and Pattern Recognition,
DOI 10.1007/978-3-662-49373-1_6

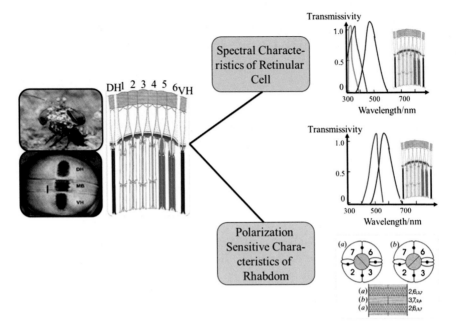

Fig. 6.1 The structure block of the multi-band vision structure of mantis shrimp

homochromatic, polarization-sensitive photoreceptors orthogonal to each other [12]. The differences of ommatidium structure show the diverse function of polarization vision such as the analysis of polarization features to distinguish the targets of interest [13] and the application of special optical features in tissue to communicate with others [14]. Research on mantis shrimp has found that spectral perception ability of the rhabdom and retinular cell is from 300 to 700 nm with perception bandwidth of 30–60 nm as in Fig. 6.1. Different types of rhabdom have different polarization sensitive ability. And the special polarization coding ability in the optic nerve systems make it possible for mantis shrimp to acquire polarization information in high resolution [4, 6, 15].

6.1.2   Compound Eye

A compound eye of insects demonstrates superior imaging performance, high sensitivity, and large field of view of nearly 360°. Figure 6.2 shows examples of insect eyes. A compound eye imaging system can expand the human vision with the inspiration of the compound eyes of insects [16].

Compound eyes are composed of parallel type and overlapped type according to different imaging principles, and the optical designs are composed of planar type

(a)   **(b)**   **(c)**

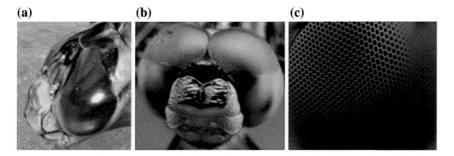

Fig. 6.2 The compound eye of a dragonfly nymph (**a**), dragonfly (**b**) and their compound eyes and microscopic images (**c**)

and curved type [17]. Rhabdom in parallel compound eye can receive light from just one ommatidium while the overlapped compound eye can receive light from several ommatidia. Table 6.1 shows some important compound eye imaging systems developed according the structure and imaging principle.

Table 6.1 Comparison of compound eye imaging system

Optical design	Principle	Type	Key techniques	Features and applications
Planar	Parallel	TOMBO [28]	Light capture component: micro-lens. Receive component:light detector. Light separation: hole array	Color imaging [28], multi-spectral imaging [29], 3Dimagery capture [30], fingerprint recognition [31]
		APCO [32]	Micro-lens array and light sensing array are installed in both side of glass substrate to connect with light sensing array	More compact structure than TOMBO, Less light interference among neighbor micro-lens
		MEMS [33]	Manufacture a 9 × 9 micro-lens array using photoresist reflow method in the glass substrate	Every lens can efficiently image an object
	Overlapped	"Cluster eyes" structure [34]	Connect focusing array, field array and relay array with light sensing array	Have larger FoV than ACPO. Capture the upright image

(continued)

Table 6.1 (continued)

Optical design	Principle	Type	Key techniques	Features and applications
Curved	Parallel	3D micro-prism structure [35]	Manufacture the 2-D micro-prism composite film into the 3D structure using micro-scale replication, stamping and casting process	Eliminate light effects among neighbor channel. Capture multi-angle images
		Self-calibration compound eye system [33]	Artificial ommatidia: synthetic rubber micro lens, photosensitive polymer waveguide and light detector	Solve the install error
	Overlapped	Curved imaging system [36]	Establish imaging model using light tools. Manufacture curved compound eye lens array using laser direct writing method	Add FOV of imaging system. Improve quality of edge imaging

6.2 Bio-inspired Multi-band Polarization Vision Model

6.2.1 Multi-band Polarization Vision Model

Dragonfly is known to have polarization vision and color vision [18] from a behavior analysis and anatomical analysis and of compound eye and ommatidia eye. A traditional compound eye model is isomorphic, each ommatidium has identical functions. In Fig. 6.3, visual images captured by different ommatidia have the same information in the common field of view, image of large field of view can be obtained by alignment and mosaicking technology. As for the vision structure of dragonfly nymph or mantis shrimp, different ommatidia have different structures and functions, each ommatidium captures image of different spectral bands or polarization angles. An isomorphic compound model is unable to describe multi-band polarization vision.

A heterogeneous compound eye structure shown in Fig. 6.4 is proposed in our work based on the characteristics of multi-band polarization vision system. Every ommatidium can discriminate band information and polarization information perceived by the compound eye. Therefore, every ommatidium is sensitive to the signal in one band or one polarization angle which will be processed to acquire a sequence of multi-band polarization images. A heterogeneous compound eye structure provides the information with the help of large FOV, and then achieves

(a)   (b)   (c)

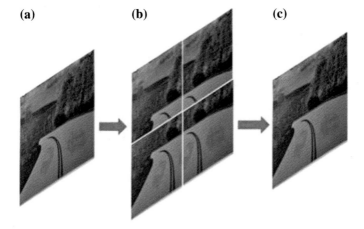

Fig. 6.3 Reconstructing a large FOV image using traditional isomorphism compound. **a** An original scene, **b** measured scene of isomorphism compound, **c** reconstructed image with increased FOV

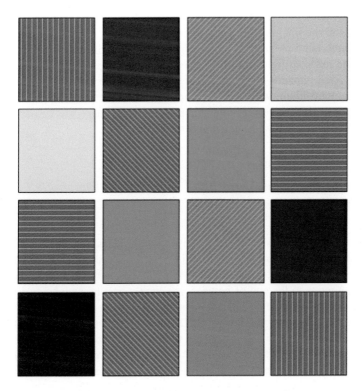

Fig. 6.4 Simulated polarization structure and heterogeneous compound eye

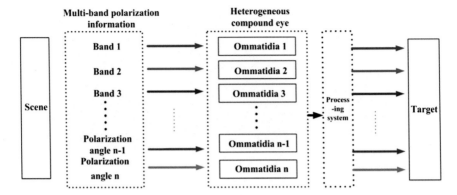

Fig. 6.5 Multi-band polarization vision model

data acquisition and target detection. In Fig. 6.5, the images of different polarization angles are firstly captured, and then the images captured by ommatidia which contain specific information in the overlapped regions, after that multi-band polarization image of large field of view can be reconstructed. Figure 6.5 shows the multi-band polarization vision model is established based on heterogeneous compound eye and the imaging perception principles.

In the multi-band polarization vision model, the polarization information contained in the reflected light from the target can be received by the compound eye system to capture images in every isolate spectral band or polarization angle with the help of ommatidium with specific functions. Then the multi-band polarization images of larger field of view can be obtained. Then the image will be processed to detect the differences between target and ground.

6.2.2   Heterogeneous Multi-band Polarization Imaging CCD Array Systems

In the compound eye vision system, only several CCDs are needed to achieve multi-band polarization imaging with an increased FOV due to their higher resolution than ommatidia. Compound eye is a highly parallel information processing system from retina to the end of nerve. It is a highly parallel and dense distributed network with adaptability, self-organization and fault-tolerant abilities [19]. Polarizers and filters are used in a CCD array to imitate polarization-sensitive microvilli and band sensitive structures in retinal cells, respectively. The difference is the filtering order between the system and compound eyes of dragonfly nymphs. Light goes through from refractive device and then be selected until arrives at the bottom of visual channel in the compound eye of dragonfly nymphs. The processing of spectral and polarization selection occurs before imaging on the CCD focal plane in the multi-band polarization system with the help of polarizer and

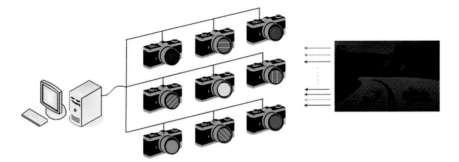

Fig. 6.6 Multi-band polarization imaging system with heterogeneous CCD array

filter. Figure 6.6 shows a multi-band polarization imaging system with heterogeneous CCD array.

The image data captured by the multi-band polarization system are not complete since there exist uncovered regions among the FOV of bands and polarization images. Therefore, an efficient data estimation method should be used to recover the multi-band polarization data cube with an increased FOV based on the correlation among images.

6.3   Multi-band Polarization Vision Systems

This section establishes the multi-band polarization vision system [20] with heterogeneous CCD array based on the multi-band polarization model in Sect. 6.2. Corresponding hardware structure and the prototype are established based on the imaging structure and the characteristics of image acquisition, transmission, and storage. The data captured by prototype is used to calibrate and register with the help of related authoring software.

6.3.1   Design of Multi-band Polarization Vision Systems

Inspired by biological vision models, we build a multi-band polarization imaging system using an array of cameras. Since the field of view and resolution of a CCD camera array is greater than a single camera, a multiple CCD camera system simulates a compound eye of aquatic insects. Since a CCD camera cannot measure polarization and spectral information, a polarizer is used to simulate the microvilli to obtain polarization information. Color filters are used to simulate the spectral selective structure of an insect eye to obtain the spectral information. Incoming light is filtered by polarizer or spectral filter and then imaged by CCD camera. There are nine CCD cameras in the proposed system, arranged in a 3 × 3 rectangular array.

Fig. 6.7 Prototype of
bio-inspired multi-band
polarization imager

Fig. 6.7 Prototype of
bio-inspired multi-band
polarization imager

Four polarizers of different angles (0°, 45°, 90° and 135°) and spectral filters of five
different bands (red, green, blue, yellow, and orange) are mounted on the camera
lens as in Fig. 6.5. Figure 6.7 shows hardware structure of the proposed bio-inspired
multi-band polarization imager.

An industrial CCD camera (Basler Ace 1300, 30 gm) of dimension
42 mm(W) × 29 mm(D) × 29 mm(H) was used for each component camera. A large
amount of image data taken using nine CCD cameras is processed and transmitted
in parallel by 9-channel Gigabit Ethernet. An optical lens (Computer M1614-MP)
with the focal length of 16 mm, viewing angles of 38° (diagonal), 30.8° (horizontal), and 23.4° (vertical) was used in all the cameras. For a cooling purpose,
CCD cameras are arranged with a gap of 22 mm.

(A) The design on the spatial location of cameras

The correlation among images of different spectral bands and polarization angles
would be exploited when reconstructing multi-band polarization image of a large
field of view. Common field of view exists in the images captured by different
cameras of the camera array. Therefore it is necessary to have proper spatial
location of nine cameras in one platform. Take the horizontal view as an example,
as shown in Fig. 6.8. Take the left view as an example, as the field of view of
camera is fixed, L is the distance between the optical axes of two cameras, because
the angle of view of camera is constant (i.e., α and β is a constant, and $\alpha > \beta$). From
Fig. 6.8, we have the following relationship

$$L = H\left(\frac{1}{\tan \beta} - \frac{1}{\tan \alpha}\right) \tag{6.1}$$

Fig. 6.8 Illustration of horizontal imaging of binocular camera

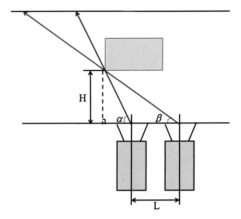

where H denotes the distance between the target and the lens. In order to increase common FOV, L should be decreased as small as possible. Same principles apply to the perpendicular view.

(B)   The requirement of structural stability

To enhance mechanical stability, the center-of-gravity position of camera array should be aligned to the center of the tripod. Cooling is another problem. If the neighboring cameras are arranged too close, the heat generated from the camera would be accumulated. In the proposed imaging system, the distance between cameras was set to 4.5 cm to allow sufficient air flow to cool down. The prototype multi-band polarization imager with heterogeneous CCD camera array was designed to satisfy the demands described above.

6.3.2   Multi-band Polarization Vision System Software

Image acquisition follows hardware set up. Considering a large amount of data from a multi-camera array and the transmission burden, MIL-Lite was used to set the imaging acquisition system which contains calibration and registration with related software to achieve pre-processing of image data. The senior programming library MIL-Lite from Matrox Imaging Library works under VS2008 environment. MIL-Lite is a senior programming library containing extended optimal functions (image acquisition, display, and storage) to support software development and improve productivity. MIL-Lite also includes image acquisition and display API that support the transport protocol of the Gigabit Ethernet to transmit and store a large amount of image data.

In the software interface, as in Fig. 6.9, there are nine image windows that display synchronized images from nine component cameras in real time. Control

Fig. 6.9 Interface of bionics multi-band polarization imaging system

buttons are located on the right. When one clicks "Start" button, the imaging system displays the scene on the left windows. With "Store" button, multi-band polarization image data captured by the nine cameras are stored on the computer simultaneously. "Stop" button stops and "End" button terminates the operating software program. The captured images tend to suffer from misalignment during the acquisition procedure, a multi-camera calibration method was come up with to obtain the internal and external parameters of the camera array and calibrate image distortion. And then SURF method was used to register multiple images of the same scene in multi-band polarization imaging.

6.4   Spectral and Polarization Information Recovery for FOV Expansion

From the geometry of the multi-band polarization imager [20] described above, each CCD camera was fixed at a spatial location that results in the disparity in the viewing angles of each camera. Therefore, a portion of spectral or polarization information would be lost in the boundary region causing a reduction in the FOV. To expand the FOV, we estimate the missing spectral and polarization information in the non-overlapping FOV region using the low-rank matrix recovery method.

6.4.1   Low-Rank Matrix Recovery

Principal component analysis (PCA) [21] has been a powerful modeling tool to achieve dimensionality reduction based on a common assumption that the measured data lies on a low-dimensional linear subspace. Although effective in the presence of small Gaussian noise in the data, classical PCA is highly sensitive to even sparse errors of high magnitude. An alternative approach to the correction of large errors in the structured data has been proposed to formulate the problem as a matrix rank minimization problem, which can be solved using nuclear-norm minimization. Estimation of missing or corrupted data in principal component analysis has become of research interests in matrix completion [22] to reduce the sensitivity to sparse high magnitude errors of classical PCA by solving a matrix rank minimization problem.

   According to the low rank matrix recovery theory, the missing information in a low rank matrix can be reconstructed efficiently via exploiting the redundancy and correlation. A low-rank matrix refers to an array of numbers having a strong correlation among its rows or columns. The rank of a matrix indicates the amount of correlation of the data. For highly correlated data, the rank of a matrix is much lower than the dimensionality of the matrix. Given the measurement $\mathbf{D}$, when the rank of data matrix $\mathbf{E}$ of the original scene is low, if matrix $\mathbf{A}$ of the missing or corrupted data is sparse compared with matrix $\mathbf{E}$, the underlying low rank matrix $\mathbf{E}$ can be recovered using the low-rank matrix recovery method [23]. Suppose that the measured data $\mathbf{D}$ can be represented by the sum of $\mathbf{E}$ and the missing information $\mathbf{A}$, $\mathbf{D} = \mathbf{E} + \mathbf{A}$, then $\mathbf{E}$ can be found by solving a constrained optimization problem:

$$\{\mathbf{A}, \mathbf{E}\} = \underset{\mathbf{A}, \mathbf{E}}{\arg\min} \{\lambda \|\mathbf{A}\|_0 + \text{rank}(\mathbf{E})\} \quad \text{s.t.} \quad \mathbf{D} = \mathbf{E} + \mathbf{A} \qquad (6.2)$$

where $\|\cdot\|_1$ and $rank(\cdot)$ denote $\mathbf{L}_0$ norm and the rank of a matrix. This means that the missing or corrupted data $\mathbf{A}$ should be sparse, only a few elements of matrix $\mathbf{A}$ are non-zero. Parameter λ balances the trade-off between sparsity and a low rank regularization. Since the $\mathbf{L}_0$ norm and the rank are both non-convex [24], solving (6.1) is an NP-hard problem, therefore we relax the criterion function by replacing the L_0-norm of $\mathbf{A}$ and the rank of $\mathbf{E}$ by the $\mathbf{L}_1$ norm and the nuclear norm, respectively:

$$\{\mathbf{A}, \mathbf{E}\} = \underset{\mathbf{A}, \mathbf{E}}{\arg\min} \{\lambda \|\mathbf{A}\|_1 + \|\mathbf{E}\|_*\} \quad \text{s.t.} \quad \mathbf{D} = \mathbf{E} + \mathbf{A} \qquad (6.3)$$

where $\|\cdot\|_1$ and $\|\cdot\|_*$ denote $\mathbf{L}_1$ norm and the nuclear norm, which are the sum of absolute values of each element and the sum of the singular values, respectively [24].

To solve the constrained optimization problem, we convert the criterion function to an unconstrained optimization version:

$$\{\mathbf{A}, \mathbf{E}\} = \underset{\mathbf{A}, \mathbf{E}}{\arg\min} \left\{ \lambda \|\mathbf{A}\|_1 + \|\mathbf{E}\|_* + \gamma \|\mathbf{D} - (\mathbf{E} + \mathbf{A})\|_2^2 \right\} \tag{6.4}$$

which can be solved using the augmented Lagrange multiplier method [25]. For a proper parameter γ, the solution to (6.4) converges to the solution to (6.3) [26].

6.4.2   Estimation of Missing Spectral and Polarization Information

In multi-band polarization imaging, the measured image data from each component camera show high redundancy and correlation among different spectral bands and polarization angles, which satisfies the low-rank matrix assumption [27]. Since the component CCD cameras are mounted in different spatial positions, a mismatch in viewing angles of each component camera causes information loss in the expanded FOV. Let $\mathbf{y}_i$ be an $m_i \times n_i$ image captured by ith component camera in either a spectral band or a polarization angle. And let $\mathbf{w}_i$ be an $\bar{m}_i \times \bar{n}_i$ image corresponding to information loss, where $m_i + \bar{m}_i = m$ and $n_i + \bar{n}_i = n$. After image calibration and alignment of all the component images, the missing parts are filled with zeroes to extend the images $\mathbf{y}_i$ and $\tilde{\mathbf{w}}_i$ to images $\tilde{\mathbf{y}}_i$ and $\tilde{\mathbf{w}}_i$ of size $m \times n$. Then a multi-band polarization image $\mathbf{x}_i$ of expanded FOV can be expressed as the sum of $\tilde{\mathbf{y}}_i$ and $\tilde{\mathbf{w}}_i$:

$$\mathbf{x}_i = \tilde{\mathbf{y}}_i + \tilde{\mathbf{w}}_i, \quad i = 1, \ldots, N \tag{6.5}$$

Figure 6.10 shows the acquisition of multi-band polarization images using a 2-D array of CCD cameras in different spectral bands and polarization angles. The CCD cameras produce the images of different FOVs. The portion of image with missing spectral and polarization information is marked as the black region in the measured data $\tilde{\mathbf{y}}_i$.

We rearrange an m-by-n image into a column vector of size mn-by-1, then form a data matrix $\mathbf{E}$ of all the vectorized image data in multiple spectral bands or in different polarization angles from N component cameras. After vectorization of $\mathbf{x}_i$, $\tilde{\mathbf{y}}_i$, and $\tilde{\mathbf{w}}_i$, (6.5) can be written in a matrix form of size mn-by-N:

$$[\mathbf{x}_1 \quad \mathbf{x}_2 \quad \ldots \quad \mathbf{x}_N] = [\tilde{\mathbf{y}}_1 \quad \tilde{\mathbf{y}}_2 \quad \ldots \quad \tilde{\mathbf{y}}_N] + [\tilde{\mathbf{w}}_1 \quad \tilde{\mathbf{w}}_2 \quad \ldots \quad \tilde{\mathbf{w}}_N] \tag{6.6}$$

We estimate the lost information $\tilde{\mathbf{w}}_i$ to reconstruct the spectral and polarization information of the image in the expanded FOV. The FOV of the reconstructed

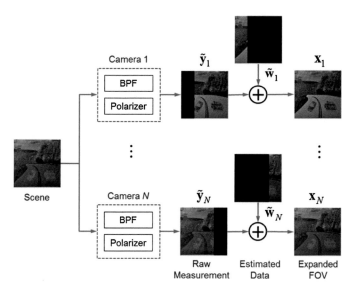

Fig. 6.10 Illustration of missing information recovery for FOV expansion in multi-band polarization imager

images $\mathbf{x}_i$ is greater than that of the measured images after calibration and alignment. A low rank matrix recovery model is given by

$$\mathbf{D} = \mathbf{E} + \mathbf{A} \qquad (6.7)$$

where $\mathbf{D} = \begin{bmatrix} \tilde{\mathbf{y}}_1 & \cdots & \tilde{\mathbf{y}}_N \end{bmatrix}$, $\mathbf{E} = \begin{bmatrix} \mathbf{x}_1 & \cdots & \mathbf{x}_N \end{bmatrix}$, and $\mathbf{A} = \begin{bmatrix} -\tilde{\mathbf{w}}_1 & \cdots & -\tilde{\mathbf{w}}_N \end{bmatrix}$.

Since multi-band polarization images are captured from the same scene in neighboring spectral bands and polarization angles, there exists high redundancy and correlation among the images in different spectral bands and polarization angles [27]. The rank of matrix $\mathbf{E}$ tends to be low. Suppose that the expanded FOV region is smaller compared to the common FOV, matrix $\mathbf{A}$ is sparse with only a small number of non-zero elements. Then the information recovery problem in the expanded FOV can be solved using the low-rank matrix recovery method:

$$\mathbf{A}^* = \arg\min_{\mathbf{A},\mathbf{E}} \left\{ \lambda \|\mathbf{A}\|_1 + \|\mathbf{E}\|_* + \gamma \|\mathbf{D} - (\mathbf{E} + \mathbf{A})\|_2^2 \right\} \quad \text{s.t.} \quad \mathbf{P}_\Omega \mathbf{A} \qquad (6.8)$$

where $\mathbf{P}_\Omega \mathbf{A} = 0$ denotes the projection of matrix $\mathbf{A}$ onto the common FOV Ω. The elements in Ω are retained, while the elements out of the common FOV are set to zero. Since the information in the common FOV can be actually measured, the elements of matrix $\mathbf{A}$ projected onto the common FOV Ω should be zero, with the constraint $\mathbf{P}_\Omega \mathbf{A} = 0$.

6.4.3   FOV Expansion with Estimated Spectral and Polarization Information

Since there are two variables $\mathbf{A}$ and $\mathbf{E}$ to be determined, we attempt to solve (6.7) by fixing one variable at a time. We solve each sub-problem iteratively until the convergence conditions are satisfied. Solving for $\mathbf{A}$ with fixed $\mathbf{E}$,

$$\mathbf{E} = \arg\min_{\mathbf{E}} \left\{ \|\mathbf{E}\|_* + \gamma\|\mathbf{D} - (\mathbf{E} + \mathbf{A})\|_2^2 \right\} \quad \text{s.t.} \quad \mathbf{P}_\Omega\mathbf{A} = 0 \qquad (6.9)$$

This sub-problem can be solved by applying soft-threshold shrinkage to the singular value of matrix $\mathbf{D} - \mathbf{A}$. The result of singular value decomposition on matrix $\mathbf{D} - \mathbf{A}$ is $(\mathbf{U}, \mathbf{V}, \mathbf{W}) = \text{SVD}(\mathbf{D} - \mathbf{A})$, where matrix $\mathbf{V}$ is a diagonal matrix, with diagonal elements of the singular values of matrix $\mathbf{D} - \mathbf{A}$. The solution to (6.8) is given by

$$\mathbf{E} = \mathbf{U}S_{\varepsilon_1}(\mathbf{V})\mathbf{W} \qquad (6.10)$$

$$S_\varepsilon(\mathbf{V}) = \begin{cases} \mathbf{V} - \varepsilon & \text{if } \|\mathbf{V} - \varepsilon\| > 0 \\ \mathbf{V} + \varepsilon & \text{if } \|\mathbf{V} + \varepsilon\| < 0 \\ 0 & \text{otherwise} \end{cases} \qquad (6.11)$$

where $S_\varepsilon(\mathbf{V})$ denotes an element-wise soft-threshold shrinkage function of matrix $\mathbf{V}$. The shrinkage threshold $\varepsilon_1 = 1/(2\gamma)$. Solving for $\mathbf{A}$ with fixed $\mathbf{E}$

$$\mathbf{A} = \arg\min_{\mathbf{A}} \left\{ \lambda\|\mathbf{A}\|_1 + \gamma\|\mathbf{D} - (\mathbf{E} + \mathbf{A})\|_2^2 \right\} \quad \text{such that} \quad \mathbf{P}_\Omega\mathbf{A} = 0 \qquad (6.12)$$

The solution to (6.12) is given by

$$\mathbf{A} = \mathbf{P}_\Omega S_{\varepsilon_2}(\mathbf{D} - \mathbf{E}) \qquad (6.13)$$

Applying element-wise soft threshold shrinkage to $(\mathbf{D} - \mathbf{E})$ with threshold $\varepsilon_2 = \lambda/2\gamma$, the solution to (6.12) is obtained by projecting the shrinkage result $S_\varepsilon(\mathbf{D} - \mathbf{E})$ onto Ω. Then we repeat the optimization process of the above sub-problems until a stopping criterion is satisfied. The algorithm stops when the objective function values do not change significantly in consecutive iterations, i.e., the norm of difference between the solutions of two consecutive iterations is smaller than a predefined constant. After the algorithm converges, the solution $\mathbf{E}$ becomes the reconstructed multi-band polarization images of expanded FOV. The FOV expansion algorithm using the low-rank matrix recovery method is summarized in follows.

Algorithm: FOV expansion using the low-rank matrix recovery method

Input: Measurement **D**
Output: Low rank data **E**
Initialization: $\mathbf{A}^0 = 0$, λ, γ, iteration count $k = 0$
While not converged Do

1. Applying SVD to $\mathbf{D} - \mathbf{A}^k$

$$(\mathbf{U}^{k+1}, \mathbf{V}^{k+1}, \mathbf{W}^{k+1}) = \text{SVD}(\mathbf{D} - \mathbf{A}^k)$$

2. Solving for **E**

$$\mathbf{E}^{k+1} = \mathbf{U}^{k+1} S_{\varepsilon_1}(\mathbf{V}^{k+1}) \mathbf{W}^{k+1}$$

3. Solving for **A**

$$\mathbf{A}^{k+1} = \mathbf{P}_\Omega S_{\varepsilon_2}(\mathbf{D} - \mathbf{E}^{k+1})$$

4. $k = k + 1$

End

6.4.4   Experiment Results

We carried out experiments to evaluate the performance of the proposed technique for estimating the spectral and polarization information in the expanded FOV. Figure 6.11 shows a test image used in the experiments that is composed of a road

Fig. 6.11 A test image with polarization objects hidden in the background

scene image in the center along with sixteen small monochrome images around it. The size of the test image is 250 mm(W) × 190 mm(H). The distance between the camera array and the test image was approximately 0.8 m. The center image consists of two parts: a color background of trees and the sky and a monochrome road surface. The monochrome part of the image contains two hidden automobile objects. The automobile objects are invisible because they have the same brightness intensity as that of the road surface. However, the automobile objects can be observed under a polarized light since they have different polarization character- istics from that of the road surface. A polarization filter or a polarized light illu- mination will reveal the difference in polarization characteristics of the hidden objects from the road background since reflected light intensity changes as the polarization angle varies. To assess the fidelity of the estimated polarization and multi-band spectral information in the expanded FOV region, we placed 18 monochrome images and small color patches around the center image. Each of the sixteen small images also contains a hidden figure having similar characteristics to

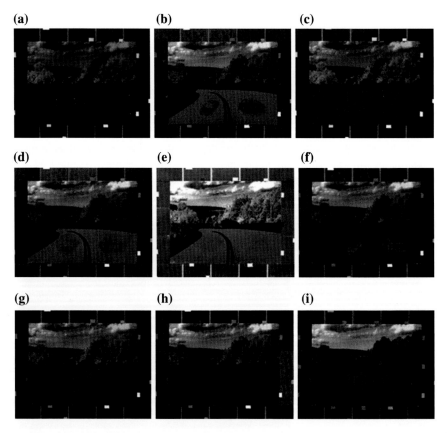

Fig. 6.12 Ground-truth spectral and polarization information of the test image. **a** red, **b** 0, **c** orange, **d** 45, **e** yellow, **f** 90, **g** green, **h** 135, **i** blue

the automobile objects in the center image. The 18 small color patches in red (4), blue (4), orange (4), green (3), and yellow (3) were pasted on the testing image in the expanded FOV region. The test image was posted on the wall to maintain the same distance from the camera.

We captured a set of pictures of the test image in different spectral bands and polarization angles. The size of the ground-truth images is 870 × 1120 in pixels. The brightness intensity was normalized to [0,1]. Figure 6.12 shows five spectral band images taken using red, orange, yellow, green, and blue color filters and four polarization images taken using polarization filters of angles of 0°, 45°, 90°, and 135°. All these pictures were taken using a single fixed camera with different color filters and polarization filters instead of a camera array to align all component images. The obtained multi-band spectral and polarization information were used as the ground truth to evaluate the performance of the estimated spectral and polarization information multi-band polarization images. Figure 6.12a, c, e, g, i denote multi-band spectral images captured using color filters of red, orange, yellow,

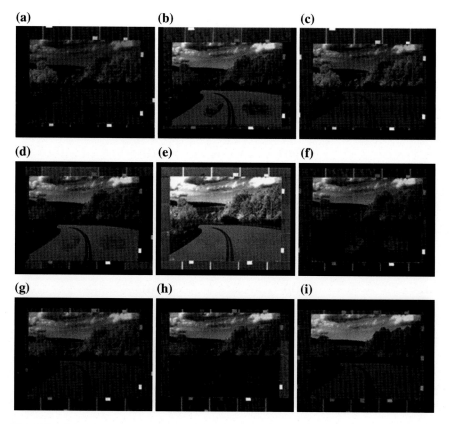

Fig. 6.13 Multi-band spectral and polarization information of the test image captured using the multi-band polarization imager. **a** red, **b** 0, **c** orange, **d** 45, **e** yellow, **f** 90, **g** green, **h** 135, **i** blue

green, and blue, respectively. Hidden automobile and woman objects could be observed in the polarization images with the angles of 0°, 45°, 90°, and 135°, as shown in Fig. 6.12b, d, f, h.

Figure 6.13 shows multi-band spectral and polarization images taken using the camera array. Due to the limited FOV, each component camera captures only a part of the scene. As a result, multi-band spectral and polarization information is lost in the extended FOV, shown as dark edges surrounding the center image. In Fig. 6.13e, the width of a dark edge corresponding to the non-overlapping FOV region in is about 80 pixels. The missing information in the dark edges in the expanded FOV can only be obtained via reconstruction algorithm. Figure 6.14 shows multi-band spectral and polarization information reconstructed using the proposed technique. The parameters λ and γ were set to 0.1 and 0.001, respectively. The dimension of the matrices **A**, **E**, and **D** was 974,400 × 9, after the minimization, the rank of matrix was 2, the rank of matrices **A** and **D** was 9. Hidden objects (a woman figure) in the expanded FOV region are visible in the reconstructed multi-band polarization images. The color patches in the expanded FOV region are

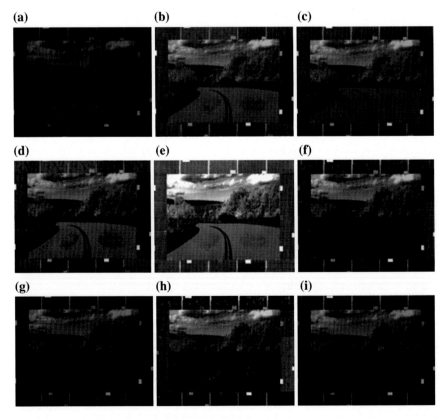

Fig. 6.14 FOV expansion with recovered spectral and polarization information. **a** red, **b** 0, **c** orange, **d** 45, **e** yellow, **f** 90, **g** green, **h** 135, **i** blue

also recovered. The FOV of the reconstructed image is expanded and is greater than any image captured by each component camera. The FOV is increased from the reconstruction method, not by changing optical lens, so the contraction between FOV and image resolution is avoided.

To evaluate the performance of the proposed reconstruction technique, we computed the fidelity of spectral and polarization information in the reconstructed multi-band polarization image, in comparison with the ground-truth data. Figure 6.15 shows the DoLP map of the ground-truth and reconstructed multi-band polarization image. The DoLP map of the reconstructed multi-band polarization image was sufficiently close to the ground-truth data. Table 6.2 lists the peak signal-to-noise ratio (PSNR), the structural similarity index measurement (SSIM), and feature similarity index measurement (FSIM) of common FOV and expanded FOV regions. The similarity index values (SSIM = 0.9985, FSIM = 0.9310) indicate that the polarization information in the expanded FOV was faithfully recovered. The polarization information in the common FOV region was well preserved (SSIM = 0.9997, FSIM = 0.9858). The similarity indices of the expanded FOV region and common FOV region indicate the accuracy of DoLP map of the reconstructed multi-band polarization images. The assessing indices of expanded FOV region are lower than indices of the common FOV region, because there is no measurement in the expanded FOV region, the decrease of assessment indices is

(a)   **(b)**

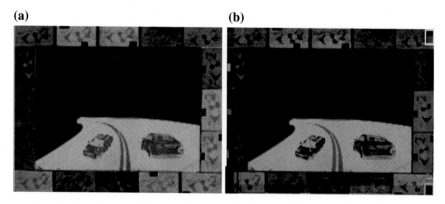

Fig. 6.15 a The ground truth of the DoLP map, **b** the DoLP map from the reconstructed polarization images

Table 6.2 The PSNR, SSIM and FSIM indices of the DoLP map in non-overlapping FOV region, common FOV region, and total FOV region

	Expanded FOV	Common FOV	Total
PSNR(dB)	14.38	23.52	18.66
SSIM	0.9985	0.9997	0.9993
FSIM	0.9310	0.9858	0.9727

Table 6.3 Average intensity and relative errors of spectral information in the non-overlapping FOV at five known spectral bands

	Red	Orange	Yellow	Green	Blue	Average
Ground truth	0.88	0.91	1.00	0.31	0.31	–
Estimated	0.74	0.71	0.89	0.26	0.22	–
Relative error (%)	15.08	21.85	10.78	16.93	27.92	18.51

Table 6.4 Viewing angles of the measured image and reconstructed image

Viewing angle	Measured image	Reconstructed image	Expansion rate (%)
Horizontal	15.77°	17.12°	8.57
Vertical	11.69°	13.05°	10.44

caused by the reconstruction error. There are some artifacts in the corner of the DOP map, as in the marked rectangle region. Since only one camera provides the information on the corner, measured data was not enough to reconstruct all the polarization information at other angles or spectral information in other bands, which led to some artifacts in the corners. Table 6.3 shows average intensity values of color patches at the five spectral bands, red, orange, yellow, green, and blue, in the expanded FOV. The accuracy of spectral information recovery was not as high enough as polarization information recovery since there are only a small the number of spectral bands and each spectral band was not narrow enough. Table 6.4 shows the amount of FOV expansion. The viewing angles were expanded to 17.12° and 13.05° from 15.77° and 11.69°, respectively, in horizontal and vertical direction, which are equivalent to an increase in FOV of 8.57 and 10.44 %.

References

1. Ibrahim I, Yuen P, Hong K, Chen T, Soori U, Jackman J, Richardson M (2012) Illumination invariance and shadow compensation via spectro-polarimetry technique. Opt Eng 51(10):107004
2. Zhao Y, Pan Q, Cheng Y (2011) Image spectropolarimetric remote sensing and application. National Defense Industry Press, Beijing
3. Barney L, Hammond JI, Sih A (2009) Dragonfly larvae and tadpole frog space use games in varied light conditions. Behav Ecol 20(1):13–21
4. Chiou TH, Allen RP (2012) A novel function for a carotenoid: astaxanthin used as a polarizer for visual signaling in a mantis shrimp. J Exp Biol 215:584–589
5. Luttbeg B, Hammond JI, Sih A (2009) Dragonfly larvae and tadpole frog space use games in varied light conditions. Behav Ecol 20(1):13–21
6. Kleinlogel S, Marshall NJ (2006) Electrophysiological evidence for linear polarization sensitivity in the compound eyes of the stomatopod crustacean *Gonodactylus chiragra*. J Exp Biol 209(21):4262–4272
7. Mathias FW, Velez MM (2012) Genetic dissection reveals two separate retinal substrates for polarization vision in drosophila. Curr Biol 22:12–20

8. Pignatelli V, Temple SE, Chiou TH, Roberts NW, Colin SP, Marshall NJ (2011) Behavioral relevance of polarization sensitivity as a target detection mechanism in cephalopods and fishes. Philos Trans Royal Soc B: Biol Sci 366(1565):734–741

9. Cronin TW, Hariyama T (2002) Spectral sensitivity in crustacean eyes. In: the crustacean nervous system, Springer, Heidelberg, pp 499–511

10. Marshall NJ, Land MF, King CA, Cronin TW (1991) The compound eyes of mantis shrimps (Crustacea, Hoplocarida, Stomatopoda). I. Compound eye structure: the detection of polarized light. Philos Trans: Biol Sci 334(1269):33–56

11. Stowasser A, Buschbeck EK (2012) Electrophysiological evidence for polarization sensitivity in the camera-type eyes of the aquatic predacious insect larva *Thermonectus marmoratus*. J Exp Biol 215(20):3577–3586

12. Labhart T, Meyer EP (1999) Detectors for polarized skylight in insects: a survey of ommatidial specializations in the dorsal rim area of the compound eye. Microsc Res Tech 47(6):368–379

13. Johnsen S, Marshall NJ, Widder EA (2011) Polarization sensitivity as a contrast enhancer in pelagic predators: lessons from in situ polarization imaging of transparent zooplankton. Philos Trans R Soc B: Biol Sci 366(1565):655–670

14. Talbot CM, Marshall JN (2011) The retinal topography of three species of coleoid cephalopod: significance for perception of polarized light. Philos Trans R Soc B: Biol Sci 366(1565):724–733

15. Thoen HH, How MJ, Chiou T-H, Marshall JN (2014) A different form of color vision in mantis shirimp. Sci 343(6169):411–413

16. Guo F, Zheng YP, Wang K (2012) Lenses matching of compound eye for target positioning. Int Soc Opt Photonics, 84200B–84200B-6

17. Gong X, Yu W, Zhang H (2013) Process in design and fabrication of artifical compound eye optical systems. Chin Opt 6(1):34

18. Bybee SM, Johnson KK (2012) All the better to see you with: a review of odonate color vision with transcriptomic insight into the odonate eye. Org Diversity Evol 12(3):241–250

19. Sarkar M Natural TA (2013) Artificial compound eye a biologically inspired CMOS image sensor. Springer, Heidelberg, pp 13–48

20. Wang Y, Yang J, Yin W, Zhang Y (2008) A new alternating minimization algorithm for total variation image reconstruction. SIAM J Imaging Sci 1(3):248–272

21. Candès EJ, Tao T (2010) The power of convex relaxation: Near-optimal matrix completion. IEEE Trans Inf Theory 56(5):2053–2080

22. Lin Z, Chen M, Ma Y (2010) The augmented lagrange multiplier method for exact recovery of corrupted low-rank matrices. University of Illinois at Urbana Champaign, Technical Report UILU-ENG-09-2215

23. Liu G, Lin Z, Yan S, Sun J, Yu Y, Ma Y (2013) Robust recovery of subspace structures by low-rank representation. IEEE Trans Pattern Anal Mach Intell 35(1):171–184

24. Candès EJ, Li X, Ma Y, Weight J (2011) Robust principal component analysis. J ACM 58 (3):11

25. Zhao Y, Yang J (2015) Hyperspectral image denoising via sparsity and low rank. IEEE Trans. Geosci and Remote Sens 53(1):296–308

26. Wang Z, Bovik AC, Sheikh HR, Simoncelli EP (2004) Image quality assessment: from error visibility to structural similarity. IEEE Trans Image Process 13(4):600–612

27. Zhang L, Zhang L, Mou X (2011) FSIM: a feature similarity index for image quality assessment. IEEE Trans Image Process 20(8):2378–2386

28. Chan WS, Lam EY, Ng MK, Mak GY (2007) Super-resolution reconstruction in a computational compound-eye imaging system. Multidimension Syst Signal Process 18 (2–3):83–101

29. Shogenji R, Kitamura Y, Yamada K, Miyatake S, Tanida J (2004) Multispectral imaging using compact compound optics. Opt Express 12(8):1643–1655

30. Yamada K, Takahashi H (2008) Small 3D image capturing system by TOMBO. In: Society of photo-optical instrumentation engineers (SPIE) conference series 6805, p 33

31. Shogenji R, Kitamura Y, Yamada K, Miyatake S, Tanida J (2004) Bimodal fingerprint capturing system based on compound-eye imaging module. Appl Opt 43(6):1355–1359
32. Radtke D, Duparré J, Zeitner UD, TŘnnermann A (2007) Laser lithographic fabrication and characterization of a spherical artificial compound eye. Opt Express 15(6):3067–3077
33. Di S, Lin H, Du R (2009) An artificial compound eyes imaging system based on MEMS technology. Int Conf Robot Biomimetics (ROBIO) 13–18
34. Duparré J, Wippermann F, Dannberg P, Reimann A (2005) Chirped arrays of refractive ellipsoidal microlenses for aberration correction under oblique incidence. Opt Express 13(26):10539–10551
35. Li L, Yi AY (2010) Development of a 3D artificial compound eye. Opt Express 18(17):18125–18137
36. Zhang H, Lu Z, Wang R (2006) Study on curved compound eye imaging system. Opt Precis Eng 14(3):346

Chapter 7
3D Reconstruction and Dehazing with Polarization Vision

Polarization vision techniques have demonstrated effectiveness in a variety of application fields including computer vision. This chapter presents 3D reconstruction and image dehazing as examples to show the benefits of polarization vision techniques.

7.1 3D Reconstruction

7.1.1 Background

Three-dimensional reconstruction attempts to extract 3D morphological information of an object at the spatial coordinates of each point on the surface. Comprehension of 3D information of a scene is greatly helpful in navigation and manipulation of military and civilian applications. 3D reconstruction techniques attempt to extract 3D depth information of an object from the captured 2D images. 3D reconstruction promotes wider applications in the fields of aerospace, shipbuilding, industrial measurement, and e-commerce [1]. Spatial coordinates of a point on the surface can be measured directly using 3D laser scanner or 3D coordinate measuring machines, which are expensive, slow, and limited by the size of target. Another approach is the 3D reconstruction based on machine vision, which is defined as the 3D topographical information using luminosity information (e.g. luminance or color) in a specific illuminate condition. Such economical and highly automatic method is most commonly used at present, which is classified as active and passive forms [2]. Passive form uses natural light as the light source without any additional active sources. Various approaches include binocular stereo vision method (as shown in Fig. 7.1) [3], shape from shading method [4], photometric stereo method [5], shape from focus method [6], motion method [7], and other methods which recover texture information.

© National Defense Industry Press, Beijing and Springer-Verlag Berlin Heidelberg 2016
Y. Zhao et al., *Multi-band Polarization Imaging and Applications*,
Advances in Computer Vision and Pattern Recognition,
DOI 10.1007/978-3-662-49373-1_7

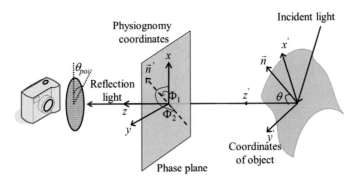

Fig. 7.1 The principle of reflection polarization imaging

1. Binocular stereo vision

Binocular stereo vision is a most commonly used passive form. Image pairs of the target are obtained by two cameras with different FOV, and the imaging point pairs of a certain point P and horizontal parallax are searched in the left image and right image using stereo matching to acquire depth information. The most prominent problem of binocular stereo vision method is the computation load of the stereo matching step, the computational complexity will increase step by step with enlarged imaging area. And the parameters of stereo vision will also influence the reconstruction results [3].

2. Shape from shading

Shape from Shading (SFD) is used to reconstruct the shape of target, which considers 3D reconstruction as the inverse processing of imaging. In this method, the grey information is closely related to the intensity of reflected light caused by 3D shape of surface, so that the radiation equation can be formulated to show the relationship between gray value and surface normal vector [4].

3. Photometric stereo

Photometric stereo method can be seen as the improvement of SFS method. An array of shading images is obtained by adjusting the situation of light source and incident angle. So the changes of pixel intensity can be utilized to recover geometric shapes on the surface [5].

4. Shape from focus

An array of images in the same scenes are acquired with different camera parameters. The sharpest focus position of each pixel will be selected to calculate the depth information of the target [6].

5. Shape from motion

An array of un-calibrated images are used to detect and match the dominant set to reconstruct the camera parameters and 3D information. The reconstruction performance depends on the intensity of dominant points, besides, this method will cost large computational load [7]. Passive forms are used to obtain the 3D topography with the help of illumination with active light sources. Although passive reconstruction methods are simple in structure, convenient in operation, and fast in data acquisition, but may not be applicable in precision measurement due to low measurement accuracy. Active forms can be classified as time modulation or spatial modulation methods. Classical time modulation utilizes the time-of-flight method which is the principle of airborne laser radar [8]. The most representative time modulation is the utilization of structured light. And the most commonly used method is the grating projection method [9]. The active method is complex in instrument, high in price, limited in measurement distance in spite of high calculation precision. It is difficult to set active light source in some practical applications.

7.1.2 3D Reconstruction with Polarization Vision

Successful results have been achieved in recent studies on 3D reconstruction based on polarization techniques. Being different from traditional machine vision methods, 3D reconstruction based on polarization vision uses the polarization information of the target surface so that there will be less effect caused by undesirable illumination conditions. The glare caused by specular reflection can be reduced at most of 50 % after polarization processing. The range of imaging is not limited because the surface texture characteristics are not necessary in polarization 3D methods. Polarization 3D reconstruction benefits from low computational cost, easy operation, low price and the capability to obtain high contrast between target and background [10].

Shape-from-polarization recovers 3D topography of an object using the polarization information of the surface of insulator [11]. The normal vector of the points on surface is calculated by the polarization information of reflection light, and then the relative depth information can be obtained by integrating. Miyazaki [12] proposed a 3D topography recovery method for transparent objects using polarization information. Atkinson and Hancock [13] proposed a 3D topography recovery method respectively using specular reflection polarization model and diffuse reflection polarization model. Huynh [14] has combined polarization with multi-spectral imaging to synchronously obtain the refraction coefficient and 3D topography of insulation. Figure 7.1 shows the principle of reflection polarization imaging. After the reflection on the surface, non-polarized incident light is reflected as polarized light. n represents the normal vector, θ represents the reflected angle (zenith angle) which is defined as the angle between n and the reflected light. Φ_1 and Φ_2 are azimuth angle which represents the angle between x-axis and the projection of n on the phase plane. Multiple groups of reflected data will be

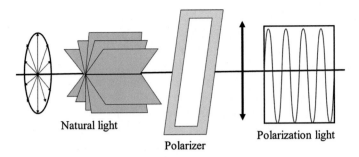

Natural light

Polarizer

Polarization light

Fig. 7.2 The principle of polarization imaging

received by rotating the linear polarizer, where θ_{pd} represents rotation angle. The azimuth angle and zenith angle can be calculated to determine the 3D shape of target.

The light can be totally polarized, non-polarized, or partially polarized. The state of polarization can be described using Stokes vectors [15] as follows:

$$\mathbf{S} = \begin{bmatrix} s_0 \\ s_1 \\ s_2 \\ s_3 \end{bmatrix} = \begin{bmatrix} \frac{1}{2}(I_0 + I_{45} + I_{90} + I_{135}) \\ I_0 - I_{90} \\ I_{45} + I_{135} \\ I_{lh} + I_{rh} \end{bmatrix} \tag{7.1}$$

where s_0 denotes the total intensity of polarized light, s_1 the intensity of linear polarization in the horizontal direction, s_2 is the intensity of linear polarization in the direction of $45°$ and s_3 represents the intensity of circularly polarization. s_3 is very small and therefore is negligible in the calculation of Stokes vector. The relative angle can be achieved by rotating the additional polarizer before the lens of camera. s_0 is the intensity sensed by camera when the polarizer is rotated. Figure 7.2 illustrates the principle of polarization imaging.

The DoLP and Orient are the most commonly used polarization parameters which can be calculated by Stokes vector [12]. DoLP represents the degree of polarization, which is also defined as the ratio of intensity of totally polarized light and the total intensity of light. The linear polarization is totally polarized so that the value of DoLP equals 1. While the DoLP of natural light is 0 and the DoLP of partially polarized light is between 0 and 1. Orient represents the angle between polarization direction and the reference direction. The standard form of normal vector is given by

$$\mathbf{N} = \begin{pmatrix} p \\ q \\ 1 \end{pmatrix} = \begin{pmatrix} \tan\theta\,\cos\varphi \\ \tan\theta\,\sin\varphi \\ 1 \end{pmatrix} \tag{7.2}$$

As for the specular reflection model, the relationship between DoLP and θ can be derived by Fresnel model [16]:

$$DoLP = \frac{2 \sin^2 \theta \sqrt{n^2 - \sin^2 \theta} - n^2 \sin^2 \theta - \sin^4 \theta}{n^2 - \sin^2 \theta - n^2 \sin^2 \theta - 2 \sin^4 \theta} \tag{7.3}$$

where n is the refractive index. Thus, θ can be derived when DoLP is known. Furthermore, Morel et al. has derived the relationship of DoLP-θ for metals based on (7.3) [17]

$$DoLp = \frac{2n \tan \theta \sin \theta}{\tan^2 \theta \sin^2 + |\hat{n}|^2} \tag{7.4}$$

where $\hat{n}$ is the complex refractive index of metal while $\hat{n} = n + i\kappa$. K is the attenuation coefficient. The value of φ is determined by Orient:

$$\varphi = \text{Orient} \pm \pi/2 \tag{7.5}$$

The gradient information which is derived from normal vector should be integral to get the relative depth or height of surface. Another form of normal vector at the observation point is shown as follows:

$$\mathbf{N} = \begin{pmatrix} -\frac{\partial Z(X,Y)}{\partial X} \\ -\frac{\partial Z(X,Y)}{\partial Y} \\ 1 \end{pmatrix} \tag{7.6}$$

Assuming that the surface function is denoted as Z(x, y), which is perpendicularly projected onto the region s_1 in x-y plane, and the gradient of any discrete point (x, y) in s_1 can be described as:

$$p(x,y) = \frac{\partial Z(x,y)}{\partial x}$$
$$q(x,y) = \frac{\partial Z(x,y)}{\partial y} \tag{7.7}$$

The integral methods can be divided as local integral and global integral. The local integral is defined as the transmission of gradient height in the gradient domain from the beginning of an initial gradient. It is strongly relative to the accuracy of data so that the nosy image data is not preferable. The results of global integral are smoother. However, global integral is sensitive to mutational directions and errors will appear at the points with low reflection index, e.g. the points at the edge of image. Shapelets integral is the most common used method which has better robustness and the capacity of edge processing [18].

7.1.3   3D Reconstruction Results

(A) Reduction of Specular Reflection

Figure 7.3 shows the experiment on the object with metal recovery. A large area of glitter is shown in Fig. 7.3a due to the strong specular reflection, which produces larger amount of saturated pixels. While the glitter suppression results shown in Fig. 7.3b prove to have good performance due to the help of polarized light.

(B) Shape Prior Based On Contours

The edge positions can be provided by contours and the normal vector of edge area is tangent to the observation direction, which means the zenith angle is approximately to be 90° (Fig. 7.4). The contours is useful for the recovery of basic shape.

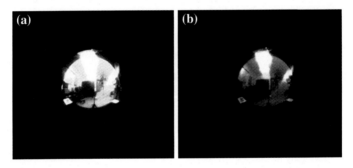

Fig. 7.3 Glitter suppression of polarization imaging. **a** Non-polarized image. **b** 0° polarization image

Fig. 7.4 The zenith angle of edge area

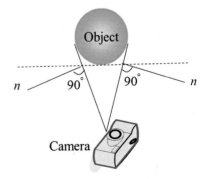

Experiments of 3D reconstruction with contours:

1. Instrument: CCD, black background, light source
2. Experimental subjects: plastic ball, rubber model
3. Experimental environment: the darkroom, single light source, the target is set in front of black background.
4. Experimental results and analysis: First extract the edge information of captured images and set the zenith angle at contours as 90° while others are 0°. Then, use Shapelet integral to obtain the basic 3D topography. The results are shown in Fig. 7.5, from the left to the right are the observation objects, the acquired contours and the recovered basic 3D topography.

Accurate 3D topography can be obtained using the basic 3D shape with the help of the normal vector information in the contours, which means that the discrete points can be used to recover the whole 3D topography with known normal vector information after leaving out the error points in contours.

1. Instrument: CCD, polarizer, dark background, light source
2. Experimental subject: simple targets (a porcelain bowl, ceramics bottle), complex targets such as an aluminum bottle
3. Instrument environment: the darkroom, single light source, set the target in front of the background

(a)

(b)

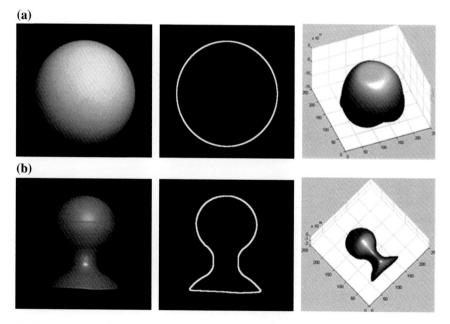

Fig. 7.5 The shape prior recovered from contours. **a** Experiment on plastic ball. **b** Experiment on rubber model

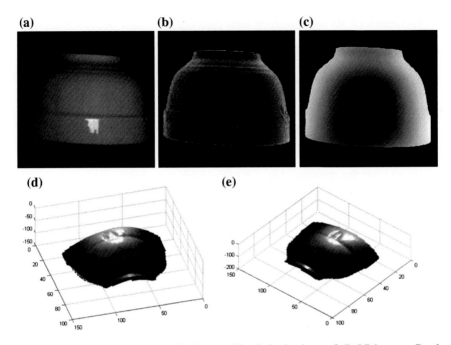

Fig. 7.6 3D reconstruction of porcelain bowl. **a** 0° polarization image. **b** DoLP image. **c** Depth image. **d** 3D topography

4. Experimental results and analysis: In Fig. 7.6, the 3D topography of porcelain bowl is well recovered. While the depression at the top is caused by the small area of glitter (Figs. 7.7 and 7.8).

The surface of complex target (such as aluminum bottle) is absorbing medium which has smaller value of DoLP than the simple targets. The reconstructed results suffer from serious noise due to limited experimental conditions.

7.2 Instant Image Dehazing

In extreme weather conditions like mist, fog, and haze, outdoor clear-imaging is particularly ill-posed due to the influence of atmospheric turbulence and suspended particles. Atmospheric particles absorption and scattering lead to the disrupting attenuation of light. The captured image data may suffer from contrast degradation, features distortion and other visual problems. To alleviate the loss of visibility, haze removal (or dehazing) is highly desired in computer vision applications, both in consumer and military fields:

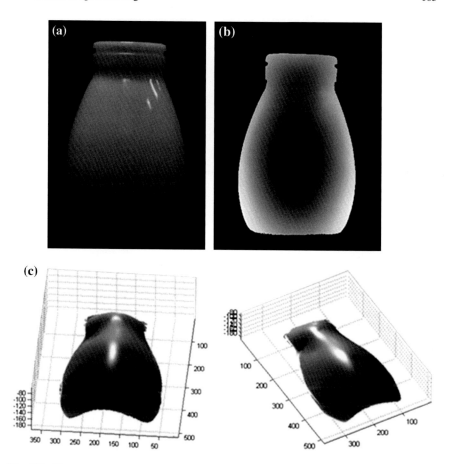

Fig. 7.7 3D reconstruction of a ceramic bottle. **a** 0° polarization image. **b** Depth image. **c** 3D topography

1. Remote Sensing: As the performance of remote target recognition will be influenced by scattering particles, so some promote works are needed to improve the results of super resolution and spectral interpretation.
2. Traffic Intelligence: The risk of traffic accidents increases in the hazy weather due to reduced visibility. Effective dehazing techniques help assist driving to improve safety [19, 20].
3. Military Engineering: Aircrafts may encounter challenges in object detection and navigation, especially in dense hazy condition. Real-time dehazing is of great significance to enhance the visibility and to improve the recognition performance.

Dehazing techniques based on image processing have been proposed to cancel out the haze effects and to obtain visually pleasing dehaze results [19, 21–25]. However, these approaches may be invalid in some special occasions due to the

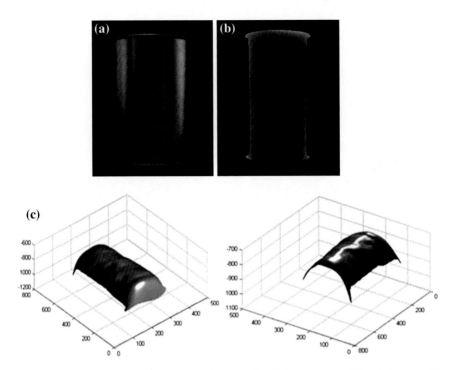

Fig. 7.8 3D reconstruction of an aluminum bottle. **a** 0° polarization image. **b** Depth image. **c** 3D topography

strong constraints and assumptions they are based on [20]. Polarization-based dehazing has been proved to be efficient in analyzing haze phenomena [26–31]. The true scenes can be reliably restored in the physics-based process. Instant image dehazing method based on polarization will be introduced in detail and the performance of haze removal will also be identified in this section.

7.2.1   Physical Models and Properties

(A)   Airlight Model

We can observe that the light intensity is gradually enhanced as the depth increased along the direction of horizon, which is referred as the airlight phenomena in related literature. Airlight phenomena is mainly caused by suspended particles backscattering and considered as one of factors for imaging distortion.

As dashed lines shown in Fig. 7.9, airlight is composed of the light come from light source and the light scattered by the atmospheric medium towards the sensor. While the contribution of backscattering light (airlight intensity) can be described as follows:

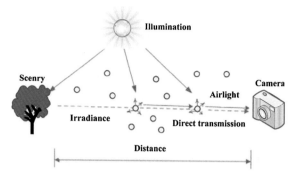

Fig. 7.9 The physical model of image formation

Fig. 7.10 Dehazing result based on polarization. **a** Observed image at vertical orientation. **b** Observed image at parallel orientation. **c** Dehazed image

$$A = A_\infty[1 - e^{-\beta(\lambda)z}] \tag{7.8}$$

where z is the scene depth, A_∞ is the intensity of airlight at an infinite distance which may be taken as the horizon. It depends on the haze density and illumination condition. $\beta(\lambda)$ is the atmospheric attenuation coefficient related with wavelength.

(B) Direct Transmission Model

In addition to the airlight component, the light emanating from the scene object also degrades the captured images due to the atmospheric particles absorption and fore-scattering. Partial light energy is scattered to other directions during the light transmission in the atmospheric medium. According to the principle of exponential decay, the light will be partly attenuated before arriving at the viewer and the attenuated signal is referred as direct transmission which is shown as solid lines in Fig. 7.11. Denoted L^{object} as original object irradiation, and the direct transmission to the camera is

$$T = L^{object} e^{-\beta(\lambda)z} \tag{7.9}$$

where z, $\beta(\lambda)$ are defined as the same meaning as in the previous part. Contrary to the airlight, the direct transmission signal is decreased as the depth increases due to the lack of visibility for distant objects, especially in dense haze conditions.

(C) Image Formation

Assuming there is no energy loss of sensors when capturing images, the overall image intensity is the sum of airlight and direct transmission. The formation of hazed image is as follows

Fig. 7.11 Dehazing of the images taken in thin haze. **a** Observed image $I^\parallel$, **b** Observed image $I^\perp$, **c** Dark channel, **d** Polarization dehazing

$$I^{total} = T + A \qquad (7.10)$$

The airlight component should be decoupled from the direct transmission after obtaining the degraded image. However, as the estimation of corresponding parameters is not easy, the acquisition of two unknowns from distorted resource is an ill-posed problem. In the following section, the polarization strategy will be adopted to solve the ambiguous problem while theoretical analysis and algorithm will be applied to remove the haze effects.

7.2.2 Image Dehazing Using Polarization Information

Polarization properties of scattering particles are taken into account in dehazing due to the advantages of polarization equipment which can acquire real-time polarization information. So the visibility of results can be improved by exploiting the polarization information to degraded images. The airlight is partially linearly polarized, the polarization effect of the direct transmission is assumed to be insignificant. However, this situation does not hold for scenarios with specularity, since the specularity will arouse higher polarization effect than the airlight. Nevertheless, as observed in general cases, the objects in the scene often contain only the diffuse surfaces [29]. To simplify the distortion model, the following two items are assumed:

1. The light undergoes single scattering in the atmospheric medium. As the multiple scatterings will depolarize the light, the overall measured polarization effect is reduced. In order to analyze the polarization properties of scattering particles, multiple scatterings are not considered.
2. The aerosols (particles) are sparely distributed, which means that the polarization effect of the incident rays does not vary along the line of light. Thus the polarization of airlight dominates the measured light [30].

Airlight component can be modulated by a mounted polarizer filter as airlight is partially polarized. The image intensities I^{total} changes with the variation of polarizer orientation angle.

The airlight intensity can be divided into two orthogonal components, $A^{\perp}$ and $A^{\parallel}$, which are respectively perpendicular and parallel to the incidence plane.

$$A = A^{\perp} - A^{\parallel} \qquad (7.11)$$

The DoLP of the airlight is given by

$$p\frac{A^{\perp} - A^{\parallel}}{A} \tag{7.12}$$

Two images captured by different orientation of polarizer are used to recover the scenes. The minimum intensity can be obtained as the polarizer is rotated to be parallel with incidence plane.

$$\begin{cases} I^{\parallel} = \frac{T}{2} + A^{\parallel} \\ A^{\parallel} = A(1 - p)/2 \end{cases} \tag{7.13}$$

While maximum intensity can be obtained as the polarizer is rotated to be perpendicular to the incidence plane.

$$\begin{cases} I^{\perp} = \frac{T}{2} + A^{\perp} \\ A^{\perp} = A(1 + p)/2 \end{cases} \tag{7.14}$$

where

$$I^{total} = I^{\parallel} + I^{\perp} \tag{7.15}$$

The effect of atmospheric particles needs to be reduced to enhance visibility. And the key step to reduce atmospheric effects is the DoLP estimation of the airlight. The DoLP of airlight was assumed to be known, and the intensity of airlight can be represented by:

$$A_u = \frac{I^{\perp} - I^{\parallel}}{p} \tag{7.16}$$

Direct transmission and transmission coefficient can be obtained from

$$T = I^{total} - A_u \tag{7.17}$$

$$e^{-\beta z} = 1 - A_u/A_{\infty} \tag{7.18}$$

Finally, the radiance of the original scene is:

$$L^{object} = \frac{I^{total} - A_u}{1 - A_u/A_{\infty}} \tag{7.19}$$

Color shift may be caused due to the estimation in RGB channels. Gamma correction is adopted to revise final dehazed results.

$$F(x,y)_{trans} = \frac{C(F(x,y) - F(x,y)_{min})}{F(x,y)_{max} - F(x,y)_{min}} \tag{7.20}$$

where $F(x,y)_{max}$ and $F(x,y)_{min}$ respectively represent the maximum and minimum intensity of dehazed image, C represents the scaled display level which is set to 255. However, A_∞ and p are the remained unknown variables need to be solve. As the polarization of direct transmission is negligible, the airlight dominates the polarization effect. An alternative is used to solve this problem.

$$\begin{cases} A_\infty = \lim_{z \to \infty} I^{total} = A_\infty^\perp + A_\infty^{\|} \\ p = \frac{A_\infty^\perp - A_\infty^{\|}}{A_\infty^\perp + A^{\|}} \end{cases} \tag{7.21}$$

where $A_\infty^\perp$ and $A_\infty^{\|}$ respectively represent the perpendicular and parallel components of infinite airlight intensity.

7.2.3   Experiment Results

The dehazing results were presented for real scene objects in dense haze condition. In Fig. 7.10a, b, the raw polarized images were captured at two orthogonal orientations of the polarizer, $I^{\|}$ and $I^\perp$, respectively. The dehazing result in Fig. 7.10c was instantly obtained with the help of polarization properties of the airlight.

Comparing Figs. 7.10a–c, the raw images were degraded due to the effect of dense haze. Dehazed images show higher contrast and visibility and the regions of green plants and distant glasshouse were clearly restored. The sign features and color fidelity of the billboard were also enhanced. Some metrics were used to quantitatively evaluate the recovered images, such as Information entropy E, Standard deviation D, and Average gradient G are estimated to access image quality. Table 7.1 shows the comparisons. The three criteria of dehazed result increased compared to the original values.

$$E = -\sum_{i=0}^{L-1} p_i \log_2 p_i \tag{7.22}$$

$$D = \sqrt{\frac{1}{MN} \sum_{i=1}^{M} \sum_{j=1}^{N} (Z(i,j) - \bar{Z})^2} \tag{7.23}$$

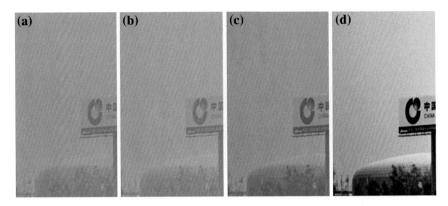

Fig. 7.12 Dehazing of the images taken in thick haze. **a** Observed image $I^{\parallel}$, **b** Observed image $I^{\perp}$, **c** Dark channel, **d** Polarization dehazing

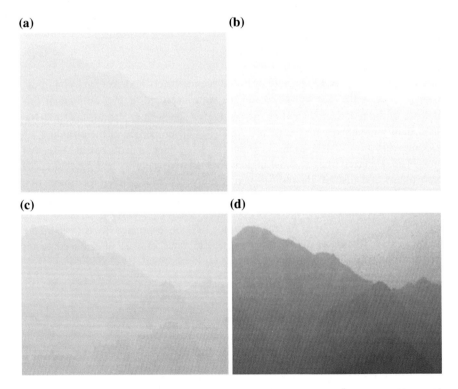

Fig. 7.13 Dehazing of the images taken in dense haze. **a** Observed image $I^{\parallel}$, **b** Observed image $I^{\perp}$, **c** Dark channel, **d** Polarization dehazing

Table 7.1 Comparisons of observed images and dehazing results

	Observed $I^{\parallel}$	Observed $I^{\perp}$	Dehazed image
E	6.3293	6.6863	11.4353
G	0.3174	0.3722	1.5053
D	7.7048	8.0433	35.0601

$$G = \frac{1}{(M-1)(N-1)} \sum_{i=1}^{M-1} \sum_{j=1}^{N-1} \sqrt{\frac{1}{2}\left\{\frac{\partial Z(i,j)^2}{\partial i} + \frac{\partial Z(i,j)^2}{\partial j}\right\}} \tag{7.24}$$

The polarization analysis of airlight can contribute to remove visual effects of haze. A database of the scenes acquired in different weather conditions is established to validate the performance of polarization dehazing approach. The results were compared to those from the dark channel prior method [2]. The weather conditions were divided into three categories: thin haze, thick haze, and dense haze according to the density of the haze. The dark channel prior and polarization dehazing methods were applied to the same scenes for comparison. Figures 7.11, 7.12, and 7.13 show dehazing results under different weather conditions.

Dehazing methods were compared based on the results acquired above. Here are the conclusions:

1. Dark Channel Prior performs well on image dehazing in terms of thin haze, but not on heavy haze.
2. Polarization dehazing demonstrates excellent results in comparison, especially for the processing of degraded images acquired in extreme weather conditions.

References

1. Chen F, Brown GM, Song M (2000) Overview of three-dimensional shape measurement using optical methods. Opt Eng 39(1):10–22
2. Mattoccia S (2011) Stereo vision: algorithms and applications, University of Bologna
3. Olson CF, Abi-Rached H, Ye M, Hendrich JP (2003) Wide-baseline stereo vision for Mars rovers. In: IEEE/RSJ international conference on intelligent robots and systems (IROS 2003) proceedings, vol 2, pp 1302–1307
4. Prados E, Camilli F, Faugeras O (2006) A viscosity solution method for shape-from-shading without image boundary data. ESAIM: Math Model Num Anal 40(2):393–412
5. Cardenas-Garcia JF, Yao HG, Zheng S (1995) 3D reconstruction of objects using stereo imaging. Opt Lasers Eng 22(3):193–213
6. Thelen A, Frey S, Hirsch S, Hering P (2009) Improvements in shape-from-focus for holographic reconstructions with regard to focus operators, neighborhood-size, and height value interpolation. IEEE Trans Image Process 18(1):151–157
7. Dellaert F, Seitz SM, Thorpe CE, Thrun S (2000) Structure from motion without correspondence. In: IEEE conference on computer vision and pattern recognition proceedings, vol 2, pp 557–564
8. Zeng Q (2009) Airborne LiDAR Point cloud data processing and 3D building reconstruction, Shanghai University

9. Zhang Y (2012) Research on three-dimensional topography measurement using structured light. Master's Thesis, Harbin Engineering University

10. Zhao Y, Pan Q (2011) Imaging spectropolarimetric remote sensing and application. National Defense Industry Press

11. Rahmann S, Canterakis N (2001) Reconstruction of specular surfaces using polarization imaging. In: IEEE conference on computer vision and pattern recognition

12. Miyazaki D, Kagesawa M, Ikeuchi K (2004) Transparent surface modeling from a pair of polarization images. IEEE Trans Patt Anal Mach Intell 26(1):73–82

13. Atkinson GA, Hancock ER (2006) Recovery of surface orientation from diffuse polarization. IEEE Trans Image Process 15(6):1653–1664

14. Huynh CP, Robles-Kelly A, Hancock E (2010) Shape and refractive index recovery from single-view polarisation images. In: IEEE conference on computer vision and pattern recognition, pp 1229–1236

15. Atkinson GA, Hancock ER (2007) Shape estimation using polarization and shading from two views. IEEE Trans Pattern Anal Mach Intell 29(11):2001–2017

16. Mahmoud AH, El-Melegy MT, Farag AA (2012) Direct method for shape recovery from polarization and shading. In: IEEE international conference on image processing, pp 1769–1772

17. Morel O, Meriaudeau F, Stolz C, Gorria P (2005) Polarization imaging applied to 3D reconstruction of specular metallic surfaces. In: International Society for Optics and Photonics. Electronic Imaging, pp 178–186

18. Kovesi P (2005) Shapelets correlated with surface normals produce surfaces. In: IEEE international conference on computer vision, vol 2, pp 994–1001

19. Tarel JP, Hautiere N (2009) Fast visibility restoration from a single color or gray level image. In: IEEE international conference on computer vision, pp 2201–2208

20. Crosby F (2004) Stokes vector component versus elementary factor performance in a target detection algorithm. In: Proceedings of SPIE 5432, pp 1–11

21. Fattal R (2008) Single image dehazing. ACM Trans Graph 27(3):72

22. He K, Sun J, Tang X (2011) Single image haze removal using dark channel prior. IEEE Trans Pattern Anal Mach Intell 33(12):2341–2353

23. Kopf J, Neubert B, Chen B, Cohen M, Cohen-Or D, Deussen O, Lischinski D (2008) Deep photo: Model-based photograph enhancement and viewing. ACM Trans Graph 27(5):116

24. Tan RT (2008) Visibility in bad weather from a single image. In: IEEE Conference on computer vision and pattern recognition, pp 1–8

25. Nishino K, Kratz L, Lombardi S (2012) Bayesian defogging. Int J Comput Vision 98 (3):263–278

26. Schechner YY, Narasimhan SG, Nayar SK (2001) Instant dehazing of images using polarization. In: IEEE international conference on computer vision and pattern recognition, vol 1, pp 325–332

27. Schechner YY, Narasimhan SG, Nayar SK (2003) Polarization-based vision through haze. Appl Opt 42(3):511–525

28. Schechner YY, Karpel N (2004) Recovering scenes by polarization analysis. IEEE Techno-ocean 3:1255–1261

29. Shwartz S, Namer E, Schechner YY (2004) Blind haze separation. In: IEEE conference on computer vision and pattern recognition, vol 2, pp 1984–1991

30. Namer E, Shwartz S, Schechner YY (2009) Skyless polarimetric calibration and visibility enhancement. Opt Express 17(2):472–493

31. Miyazaki D, Akiyama D, Baba M. Furukawa R, Hiura S, Hiura N (2013) Polarization-based dehazing using two reference objects. In: IEEE international conference on computer vision workshops, pp 852–859

Printed in the United States
By Bookmasters